James St. Grimes

Geonomy

Creation of the continents by the ocean currents. An advanced system of

physical geology and geography

James St. Grimes

Geonomy
Creation of the continents by the ocean currents. An advanced system of physical geology and geography

ISBN/EAN: 9783337316976

Printed in Europe, USA, Canada, Australia, Japan

Cover: Foto ©berggeist007 / pixelio.de

More available books at **www.hansebooks.com**

GEONOMY:

CREATION OF THE CONTINENTS

BY

THE OCEAN CURRENTS.

AN

ADVANCED SYSTEM OF PHYSICAL GEOLOGY AND GEOGRAPHY.

"The Spirit of God moved upon the face of the waters."—GENESIS.

BY

J. STANLEY GRIMES,

AUTHOR OF "PROBLEMS OF CREATION" AND "MYSTERIES OF THE HEAD AND HEART."

PHILADELPHIA:
J. B. LIPPINCOTT & CO.
1885.

PRINTED AND STEREOTYPED BY

J. B. LIPPINCOTT & CO.,

PHILADELPHIA.

Dedication.

CONTENTS.

		PAGE
INTRODUCTION		7
SYNOPSIS		23
SECTION	I.—Preliminary and Historical . . .	25
"	II.—The General Ocean Currents . . .	37
"	III.—The Elliptical Currents	52
"	IV.—Effects of Oceanic Resistance . . .	57
"	V.—Lack of Symmetry	60
"	VI.—Local and Counter-Currents . . .	63
"	VII.—Limits of the Ellipses	68
"	VIII.—Extension of the Continents . . .	71
"	IX.—Loxodromic Trends	76
"	X.—The Sediment	79
"	XI.—The Northern Glacial Epoch . . .	86
"	XII.—The Southern Glacial Epoch . . .	92
"	XIII.—The Mountain Upheavals	97
"	XIV.—The North Indian Ocean	105
"	XV.—Concluding Review	112

INTRODUCTION.

BY REV. W. R. COOVERT.

I REGARD it not only as a pleasure but a privilege to write this introduction. Like all others who have enjoyed a long and intimate acquaintance with the author, I have always had the greatest respect for his character and admiration for his abilities. But I do not propose to avail myself of this opportunity to eulogize him personally: his merits are best illustrated by his works, and I am aware that he desires no undeserved compliments. One of his most distinguishing traits is his intellectual independence.

I take especial interest in this work for the reason that it was written in compliance with my urgent request. I had long known that the author entertained peculiar views concerning the origin of continents, but my professional avocations prevented me from giving special attention to the subject; besides, I must confess that, notwithstanding my high opinion of his abilities, I supposed that his novel ideas were only ingenious speculations. When, therefore, he came to Pittsburgh, and lectured before the Teachers' Academy, he not only convinced me that he had made a discovery which would

7

produce a revolution in geology and physical geography, but I found, upon inquiry, that he had made a similar impression upon the minds of all his most competent hearers, not only in Pittsburgh, but in Grove City and Waynesburg, and several other colleges. Under these circumstances, they very naturally inquired for a book containing an account of the new "Geonomy;" and I was surprised to learn that all the books relating to the subject were out of print, excepting the "Problems of Creation," and that only contained a synopsis comprised in a few pages, the greater part of the book being devoted to other topics. When I afterwards urged my friend to prepare a more detailed account of his system, and took the liberty to intimate that he did not seem to appreciate the value of his own discoveries, I received the following characteristic reply, which, as it justifies my writing this introduction, I shall take the liberty to insert:

" CHICAGO, Nov. 1, 1884.

" REV. W. R. COOVERT.

" DEAR SIR,—In answer to your last note, I will say that, in compliance with your suggestions, I have begun to prepare a small treatise on Geonomy, which, I hope, will meet your expectations. I will send you the manuscript as soon as it is finished, and, as you seem to apprehend that I shall leave out some things that I ought to insert, I will request you to add such notes as you may think necessary, and also to write an appropriate introduction ; with the proviso, however, that you do not allow your partiality to make any unnecessary allusions to me personally. You remark that I 'appear to undervalue the importance of my own work.' You are greatly mistaken. I have not the slightest doubt that the time will come when the elliptical theory of Geonomy will be taught in all the higher schools of the civilized world. ''Tis a consummation devoutly to be wished,' but I shall not live to witness it. The history of scientific advances is instructive on this subject. The discovery of Copernicus was treated with contempt by Lord Bacon, the greatest philosopher in the world, fifty years after Copernicus died. That 'history repeats itself' is as true now as ever. The soldier who ventures too far in advance of his fellows is not only in danger from the enemy, but he will be very likely to be shot in the back by some of those whose support he had a right to expect. Do not understand me as complaining. The follower of science, like an enthusiastic hunter, finds his greatest reward in the pleasure of the pursuit. We have the authority of the 'Encyclopædia Britannica' for the statement that 'the universities of Europe were often the fastnesses from which prejudice and error were the latest in being expelled. For more than thirty years after the publication of the discoveries of Newton, the system of Descartes kept its ground in the British universities.' The principles of Newton were at length smuggled in by a stratagem ; they were introduced, without authority, in the form of notes to a new edition of the old erroneous textbook.

"Sir Charles Lyell says, 'We are sometimes tempted to ask whether the time will ever come when science

shall have attained such an ascendency in the education
of the millions, that it will be possible to welcome new
truths, instead of always looking upon them with fear
and disquiet; and to hail every important victory gained
over error, instead of resisting new discoveries long after
the evidence in their favor is conclusive.'

" Yours respectfully,

" J. STANLEY GRIMES."

This treatise is especially devoted to the elliptical
theory of the origin of continents; but the reader must
not suppose that the author has labored successfully in
no other departments of science. He has made important advances in mental physiology and in astronomy;
and, disconnected as these topics may at first seem, his investigations have tended to bring them within the sphere
of his one great theme,—creation by theistic evolution.

His first publication was made in 1838. In that work
he demonstrated that the functions of the brain as well
as of the body include three classes : those that relate to
self, those that relate to society, and those that relate to
knowledge. Each of the three classes of mental organs
may be said to have its roots in a class of analogous
bodily organs, and from thence, like the trunks of three
trees, they extend up into the brain, where they expand
and send off special branches.

In that work the author did not, directly and pro-
fessedly, advocate evolution ; he merely pointed out the
fact that the cerebral organs are arranged in a certain
progressive order, which would now be at once recog-

nized as the result of evolution. But in 1850, eight years prior to the publication of Darwin's first work on evolution, he published a volume which was more Darwinian than Darwin himself, excepting that it was decidedly and avowedly theistic; indeed, it was the first book ever published on purpose to advocate *theistic evolution*. On the title-page he placed the following sentence: "Circumstances are the fingers of God, by the agency of which he created and controls all things."

The theory advocated in that work is that all the higher species of plants and animals are derived from the lower, and that all the changes that occurred in organized beings were caused by changes in the condition of the earth during the successive geological ages.

The book was received in New England with the greatest disfavor by all parties. Its author stood alone —the solitary advocate of creation by theistic evolution—in this country if not in the world. The infidels denounced it, in the " Boston Investigator," on account of its recognition of a personal Creator, and its implied admission of the truth of Christianity; the Puritans condemned it for the reason that it was opposed to the literal interpretation of the book of Genesis. Dr. Jarvis, the historiographer of the Episcopal Church, in a letter to the author's pastor, the Rev. Orange Clark, expressed the general opinion by declaring that " A man who could seriously advocate the theory that the human brain was gradually created by receiving successive additions during the geological ages, was a fit candidate for a mad-house."

This was thirty-four years ago. If the learned historiographer could have lived to the present time, he would have been much more astonished to find theistic evolution defended by some of our most eminent divines, including Dr. McCosh, the president of Princeton College, and the Rev. Joseph Cook, of Boston. The truth is that the book was then, at least in New England, far in advance of ·the times. It has lately been reissued under the title of "Problems of Creation," by Henry A. Sumner & Co., Chicago.

The author's long-continued and thorough investigation of Mental Physiology proved to be of great service to him, as well as to the public, when Mesmerism and Modern Spiritism were agitating the country. The phenomena of Mesmerism, in various forms, have been observed in every age and almost every community, and have been attributed to witchcraft, to imagination, to magnetism, to electricity, and to the inspiration of good and of evil spirits. Some physiologists have supposed that the trance may be induced by the long-continued concentration of the subject's mind, others that it results from "expectant attention," and still others that it may be induced by gazing upon a bright object; the most prevalent theory has been that the will of the operator subdued and controlled the mind of the subject.

In 1876 our author published his "Mysteries of the Head and Heart explained," in which, for the first time, a really reliable and scientific explanation of the facts was given. He demonstrated that all the phenomena are produced by the abnormal excitement and dominance

of a particular group of mental organs, which he denom-
inated "the conforming group." It may be confidently
asserted that this book contains the only true system of
Mental Physiology ever published,—the only one that
furnishes a rational explanation of the influences of the
body and the mind upon each other. The late Dr.
Geo. M. Beard, one of the most distinguished physicians
of New York, in a lecture before a large audience, with
his usual generosity and frankness, said, "All that we
really know of the physiology of mesmerism and spiritism we have learned from J. Stanley Grimes." Hun-
dreds of physicians and others have made the same
declaration. I can speak on this subject from personal
experience. After having read everything that I could
find in the books relating to mesmerism, religious trance,
spiritism, and prayer-cures, I find my friend's theory to
be the only one that will bear a rigid scrutiny, or stand
the test of carefully-conducted experiments. When we
read the explanations of other writers—metaphysical or
medical—we find ourselves treated to conjectures, asser-
tions, and assumptions; but he confines himself strictly
to the principles of physiology, and to facts that can
readily be verified.

I can only allude briefly to our author's advances in
Physical Astronomy, and, for further details, refer to
his "Problems of Creation." His theory concerning
the origin of the sun and planets is much more consistent with the facts than is the "Nebular Hypothesis"
of Laplace. He assumes that the nebula became separated into a sun and a revolving disk; and that the dif-

ferences of orbital velocities, in the different parts of the disk, produced a separation into a series of concentric rings, which would necessarily be wider with distance from the centre, in proportion as the orbital velocities decreased. When the rings became concentrated into globes, their intervals, also, of course, increased with their distances. Astronomers have never been able to account for the well-known fact that the intervals increase; but here we have a perfectly satisfactory reason. Had the disk been of uniform thickness, the magnitudes of the planets would also have increased with distance; but a glance at a diagram representing their actual relative magnitudes informs us that the disk must have been very irregular in thickness; and this fact accounts for the irregularities, not only of the magnitudes, but also of the intervals.

It is not easy to decide which of the discoveries made by our author is of the most value; but the probability is that the one to which this book is especially devoted will ultimately occupy the highest place in the history of science. When I assert that the discovery of the laws of the elliptical currents, and their agency in the creation of the continents, bears the same relation to geology that Newton's discovery bears to astronomy, I merely state a fact that every candid and capable reader of this book will be forced to admit. Geology is usually termed "a science," and it is one in the same sense that astronomy was a science previous to the discovery of the law of gravitation. Before that time it was known that there were several planets which revolved around the sun, that they moved in elliptical orbits, and that their periods and distances were mathematically proportional; but the reason and the causal connection of these facts were utterly unknown until Newton's law revealed them. A general law, like that which explains the elliptical planetary orbits, or that which explains the elliptical ocean currents, is a chain that binds innumerable facts together into a single scientific system. These laws are the brightest stars that illuminate the celestial vault of science. A knowledge of isolated facts and minor details may be acquired by the lowest men, and even by the lower animals; but a knowledge of the general laws of nature is ennobling, and partakes of the sublime; it is the nearest approximation to Omniscience that a human mind is capable of making.* The elliptical theory produces

---

* I am unwilling to allow this opportunity to pass without calling attention to some of the coincidences between the Genesis of the Bible and of modern science.

1st. Genesis is the first and most ancient book in which mankind were informed that there was a time when this world did not exist; modern science teaches the same truth.

2d. Genesis teaches that after the world was created "the earth"—the "dry land"—was without "form and void." Geology now confirms this statement.

3d. Genesis represents the water, in the beginning, as covering the whole world; geology demonstrates that before the "dry land"—the continents—were formed, the globe was covered by the ocean.

4th. Genesis affirms that "the Spirit of God moved upon the face of the waters," gathered them together," "called them seas," and caused the "dry land"—the continents—to "appear." This is the fundamental proposition of geonomy.

even a greater revolution in geology than gravitation
did in astronomy. It banishes a greater number of
fallacies, and explains a greater number of otherwise
unaccountable facts. This becomes evident when we
compare it with its only rival, the cooling and shrinking theory, which, even if true, explains nothing,
whereas the elliptical theory, combined with geology,
explains nearly everything. Newton proved that a
combination of forces caused the *planets* to move in
elliptical paths, and our author that a combination of
forces caused the *ocean currents* to move in elliptical

5th. Genesis represents the "dry land" as not appearing until
the third day or period, and geology proves that the globe and the
ocean existed several ages before the dry land. But, according to
the "cooling and shrinking" hypothesis, which Prof. Dana, in
accordance with nearly all geologists, assumes to be an "admitted
fact," the dry land must have appeared first, for water could not
possibly have remained on a red-hot globe.

Our theological geologists must acknowledge that, if their theory is true, it is very strange that Moses neglected, or was not in-
spired, to inform us that the land was not only dry but "red-hot"
before the ocean enveloped it.

6th. Genesis represents the aquatic animals as being created
first, the higher land animals next, and man last of all. The
geological fossils confirm this representation.

In conclusion, I would remark that it is the duty of every
Christian soldier to "gird up his loins like a man," and contend
for the truth which God, in his providence, imparts to us,
whether it comes in the form of science or of Divine revelation.
In the language of Milton,—

"Let truth with error grapple.
Who ever knew her put to worse in a free and open encounter?"

W. R. C.

paths when there were no lands. If we study both of these systems of ellipses, and compare them, we find that our author had much the more difficult task to perform, and much the more complex problem to solve. The planetary ellipses are produced by two forces only, one tangential to the other, and both acting constantly; the oceanic ellipses are produced by the operation of *four* inconstant forces,—north, south, east, and west.

In the first section of the ellipse, from the twentieth parallel to the forty-fifth, the north and east forces combine; in the second section, the east force acts singly; in the third, the east and south combine; in the fourth, the south and west combine; in the fifth, the west acts singly; in the sixth, the north and west combine. To render the problem still more complex, the resistance of the inert ocean to the passage of the current through it must be taken into the account; and our author has demonstrated that its effect is to vary the form of the ellipse and make it irregularly rhomboidal. But, notwithstanding the complexity of the forces that produce and that derange the ellipses, they harmonize most wonderfully with the outlines of the continents.

It may be remarked, that the laws and forces by which the planetary ellipses are produced would remain the same if the planets were all stricken from existence; so, also, the laws and forces that produce the oceanic ellipses would exist if there had never been any oceans or continents on the globe. If Mars, or any other planet, has an ocean analogous to ours, the same forces and causes must produce analogous effects upon its currents. If there had been no sediment for the ocean currents to distribute, the three pairs of ellipses would have been formed just the same, and would have continued to circulate through all time, and with even greater regularity than they actually have done.

There are but few sciences that have a mathematical basis. Astronomy, optics, chemistry, and mechanics have this great advantage. Our author's discovery adds another to the list, and thus elevates it to the highest rank.

W. R. COOVERT.

PITTSBURGH, PA.

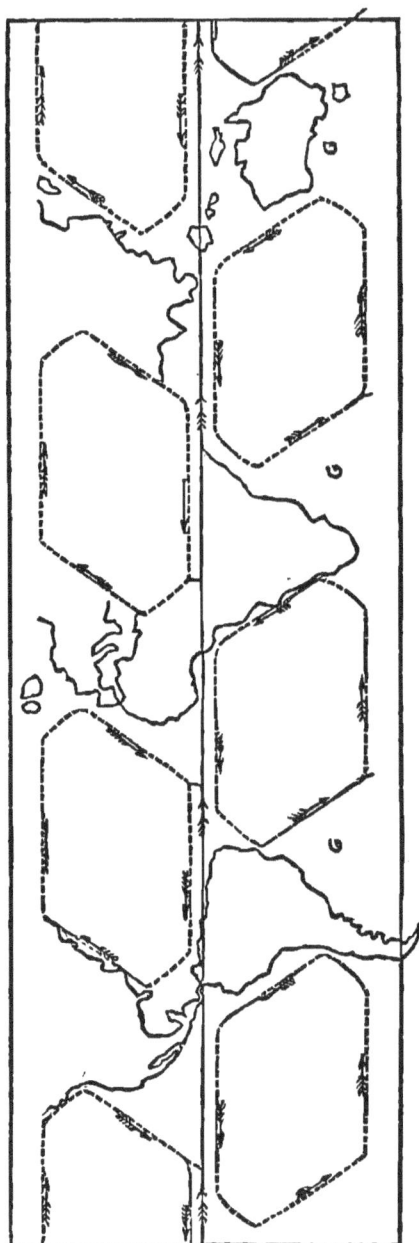

Fig. 1.—Ideal Map.

20

FIG. 1.—IDEAL MAP.

This ideal map represents and illustrates the whole of geonomy at once.

The reader will perceive that the six ellipses are drawn and arranged in strict accordance with mathematical and theoretical principles. They are all alike in size, they are of the same form, their intervals are the same, and each southern ellipse is the same distance east of its northern mate; yet with all these peculiarities, it will be observed that the outlines of the present actual continents, which are drawn in the intervals, coincide with the elliptical currents in such a manner as to prove that the relation must be one of cause and effect.

The map demonstrates that the analogies and repetitions of the forms of the continents, and also the trends of the shores, result from the forms and positions of the ellipses.

The principal departures from the elliptical ideal map have been produced by the raising of the bed of the North Indian Ocean, and by the southern Glacial Epoch, which caused the sinking of the areas marked G G G.

The arrows in the equatorial line indicate the course of the equatorial counter-current, which flows between the two westward equatorial currents. The other arrows indicate the directions of the elliptical currents.

We would call special attention to,—

1. The analogy between the western curve of southwestern North America and of southwestern Africa.

2. The analogy between the Caribbean Sea, including the Gulf of Mexico, the western part of North Africa, and the China and East Indian Sea.

3. The northeast trend of the eastern shore of the United States and of the eastern shore of Asia.

4. The hollow in the northwest coast of North America and the southwest coast of South America.

5. The hollow or gulf on the west of Panama and of the Gulf of Guinea.

6. The pointed extremity of South America at Cape St. Roque and the pointed extremity of eastern Africa.

7. The loxodromic trend of Central America and of the islands between Asia and Australia.

21

GEONOMY.

SYNOPSIS.

THAT the reader may form a general idea of the system of Geonomy without perusing the whole book, the most essential propositions are included in the following synopsis:

1. All the elevations of the earth's crust have resulted from the sinking of the ocean basins, or of smaller local basins, beneath the weight of the sediment.

2. When the ocean covered the globe, there were three pairs of elliptical currents that collected sediment on the ocean's floor, the weight of which produced three pairs of sinking basins, namely, the North and South Atlantic, Pacific, and Indian Ocean basins.

3. The fluid or plastic lava forced from beneath the sinking basins was driven under the crust in the interoceanic spaces, and by raising them created three pairs of continents, namely, North and South America, Europe and South Africa, and Asia and Australia.

4. The elliptical currents in each hemisphere were all limited to the zone between the equator and the forty-fifth parallel.

5. The southern ellipses were placed about fifty degrees of longitude east of the northern, and, consequently, the continents are equally unsymmetric; besides, the central, or tropical continents—the isthmuses—are distorted.

6. The resistance of the inert ocean to the passage of the currents caused the ellipses to be irregular in form, and these irregularities are impressed upon the outlines of the continents.

7. The lava that was forced from beneath the sinking basins up under the continents, raised those parts of the continents most that were nearest to the ocean; consequently, it is there that all the plateaus, or high table-lands, are situated. It is also there that the surfaces of the continents are most corrugated by the lava beneath the crust; the lowest lands are at a considerable distance from the great oceans.

8. While continents, including plateaus, were raised in consequence of the sinking of the great ocean basins, nearly or quite all continental islands, and upheaved mountain ranges, resulted from local and limited depressions made in submarine parts of rising continents.

9. The raising of the land between the 45th parallel and the pole, by excluding the warm tropical currents, produced glacial epochs in the circumpolar regions.

10. The floor of the North Indian basin has been raised above the sea and added to the continents of Asia, Europe, and Africa.

11. Earthquakes are produced by the perceptible movements of the lava under the continents from beneath sinking basins.

SECTION I.

PRELIMINARY AND HISTORICAL.

THE word geonomy is from the Greek *ge*, the earth, and *nomos*, a law, and is analogous to astronomy, which is from *astron*, a star, and *nomos* (see Worcester's Unabridged Dictionary). Unfortunately, there is no name in use for the science of the earth besides geography, which literally signifies a description of the earth, or geology, which is a discourse concerning the earth, and which is generally understood to relate to the successive changes of the earth before reaching its present condition. A term is wanted that, without including the details of paleontology, will be sufficiently comprehensive to embrace not only a description of the appearances which the surface of the globe presents, but also an account of the dynamical causes that produced those appearances. Hitherto, geology, with all its achievements, has failed to furnish the principles which are requisite to make physical geography a system; the two subjects have had so little in common that, in many of our schools, they have been taught separately. It is expected that this treatise, by giving reasons for the existence, the forms and positions of the continents, will supply the links necessary to connect the two subjects.

The reader will find that this is not a controversial work. None of the generally admitted *facts* of geology

B   3

are questioned, though the commonly-received hypoth-esis of "the cooling and shrinking of the globe" is no longer supposed to be needed to account for the exist-ence of oceans and continents. If it is true that the shrinking of the interior of the earth rendered it neces-sary for its crust to subside in some places, and main-tain its position in others, it is obvious that it must have subsided the most readily and rapidly in those places where it was most heavily loaded with sediment. Again, if the crust, in sinking beneath its own weight, would produce "lateral thrusts" against the borders of the continents with sufficient force to raise them and to corrugate the strata, surely the addition of stratified sediment, more than a mile in thickness, would not render the lateral pressure any less effective. It will be found that while the novelties we introduce in these pages are additions to our knowledge, they are not hos-tile to any of the reasonable doctrines formerly inculcated by competent teachers. The discovery of the forces that produce the elliptical orbits of the planets did not set aside any of the facts of astronomy previously known, but it added many new and interesting truths, and bound the whole together into a beautiful and system-atic science. Very similar, we believe, will be the consequences of the discovery of the causes and effects of the elliptical currents of the ocean upon physical geology.

When the observations and reports of the great navi-gators of the sixteenth century had enabled geographers to draw outline maps of all the continents, philosophers began to speculate concerning the causes of their peculiar forms and analogies. It was observed that North and South America constitute a pair of continents united by an isthmus; Europe and Africa, a second pair, also united by an isthmus; and Asia and Australia is a third pair, connected by a chain of islands that are equivalent to an isthmus. The three pairs seem to be repetitions of each other and to suggest the idea that the forces, whatever they were, that produced one pair, repeated their operations to produce another, and then repeated again to produce a third pair.

The continents are all broad at the north and pointed at the south, and the trends of their shores are, almost without exception, loxodromic; that is to say, they do not trend directly north and south, nor east and west, but northeast and southwest, or northwest and southeast. There is another fact that seems to have been overlooked until it was mentioned by the present writer, and that is that the analogies of the continents are all confined to the zone between the equator and the forty-fifth parallel. We shall have occasion to refer to this fact hereafter. It was also observed that while there were three pairs of continents, there were only two and a half pairs of oceans; the former existence of the North Indian was then unknown. Alexander Keith Johnson, the learned British geographer, remarked that all the continents together are equal to three Americas. President Edward Hitchcock, in a conversation with the author, said that it would be a strong recommendation of any new theory of the earth, if it would account for the analogies of the continents.

The Baron Humboldt expressed a doubt whether the
causes of the continental forms would ever be discovered.
He founded this idea upon the supposition that the forces
which upheaved the lands are in the interior of the globe
beyond the sphere of human observation.* We shall endeavor to demonstrate that the reason why so many distinguished scienticians have failed in their researches, is
because they were not in possession of the elliptical key,
without which it was impossible to unlock the mystery
of the ocean.

In 1649, before the founders of the science of geology
were born, Descartes proposed the cooling and shrinking
hypothesis of the origin of oceans and continents, which
was afterwards advocated by Leibnitz, by Buffon, and
by Cuvier, and has since been sanctioned by nearly all
geologists. They supposed that the earth was at first a
globe of fiery fluid, and that it cooled until a crust was

---

* "Very little can be empirically known concerning the causal
connection of the phenomena of the formation of continents, or of
the analogies and contrasts presented by their configuration. . . .
All that we know is that the active cause is subterranean. We
deem the force accidental owing to our inability to define it, as it
is removed from the sphere of our comprehension."—*Humboldt's
Cosmos.*

" No adequate cause has yet been assigned for the present distribution of land."—*Prof. Geikie, Edinburgh.*

" The cause of the present positions of the dry land is as yet
beyond the indications of science."—*Prof. Page, Edinburgh.*

" We are far from a rational explanation of the observed forms
of the continents. . . . The inequalities of the earth's crust are
facts in nature that have arisen from the conflict of manifold forces
acting under unknown conditions."—*Encyclopædia Britannica.*

formed upon its outer parts; ever since then the interior
has continued to cool and shrink, and the crust to sink
in some places and maintain its elevation in others.
In 1787, Werner, a mining engineer, drew attention
to the fact that the rocks in Germany are constituted and
arranged in such a manner as to indicate that they are
composed of sediment, which has been deposited in successive strata at the bottom of the sea. Shortly afterwards, William Smith, an English engineer, announced
the much more important discovery that the fossils in the
strata indicate their relative ages. This brought to the
investigation a great number of expert naturalists, whose
combined labors have created the present system of
geology and paleontology.

Although this science has not revealed the causes that
gave the continents their forms and positions, it has
pointed out the areas that first emerged from the ocean,
those that rose next, and so on to the present time. We
have learned that the first dry lands in North America
were those in the northern parts of the continent, and
that they gradually extended, first south and east, and
then west. The same is true of South America and
of Asia. Europe commenced rising in the north, and
the dry land gradually extended south and east. We
know that not far from two-thirds of Europe, Asia, and
Northern Africa were beneath the sea, when the greater
part of North America was dry land. We also know
that in the middle geologic ages, when but small areas
of the present continents were dry land, the remainder
was only covered by comparatively shallow water, and

was alternately rising and sinking, but on the whole was making its way slowly to its present elevation. But these and numerous other revelations, interesting as they are, shed no light upon the mysteries of physical and dynamical geology.

The researches of practical geologists have extended but little below the surfaces of the lands, and have been guided almost exclusively by the fossil remains. If they have attempted to grapple with the problems of dynamical geology, and to account for the elevation of continents and mountains and the depressions of ocean basins, they have begun by adopting the hypothesis that the globe, ever since its creation, has been cooling and shrinking, and then they have proceeded to infer that all the present inequalities of the earth's crust are the necessary effects of this cause. No one pretends that this hypothesis has been proved to be true, and for that reason we are not required to prove it to be false. It was *invented* upwards of two hundred years ago for the purpose of accounting for the origin of oceans and continents, and it has been retained and *assumed* to be true for no other reason than that no better has been proposed. Vice-President Hitchcock, at the meeting of the American Association at Minneapolis, 1883, in his address said,—

" As we are endeavoring to advance science, we must touch debatable topics.

" We must assume the correctness of the commonly-received opinions concerning the earliest history of our planet,—that it has passed through the condition of a burning sun, the period of igneous fluidity. By subsequent refrigeration it has become partially or wholly solid."

Prof. James Hall, of Albany, refuses to advance science by assuming the correctness of this hypothesis. He has devoted a long life to the personal examination of geological formations, and has arrived at the conclusion that "we *know nothing* about the history of the globe before the ocean covered it."

Principal Dawson, the president of the Association, in his address at Minneapolis, after admitting that "the causes and mode of operation of the great movements of the earth's crust are still involved in mystery," adds that "one potent cause is the unequal settling of the crust toward the centre; but it is not generally understood, as it should be, that the greater settlement of the ocean's bed has necessitated its pressure against the sides of the continent, in the same manner that a huge ice-floe crushes a ship in a pier.

"The rocks of Pennsylvania and Maryland (the Appalachians) have been driven back in a curve to the west."

It would be interesting to hear the learned ex-president apply his driving-back theory to the curves of the Carpathians, the Alps, the Himalaya, and still more to the curves of the Antilles and the Aleutians. The curves in those mountains and islands indicate that the *driving* has been *toward* the ocean, and "back in a curve" from the smaller basins and also *from* the continents.

In 1857 the author wrote a small book (published by Phillips & Sampson, Boston), in which he endeavored to

prove that the three pairs of continents were raised, in
consequence of the sinking of the three pairs of ocean
basins beneath the weight of the sediment distributed
by the elliptical currents. At that time he was unable
to give a satisfactory explanation of the causes of those
currents; his work was therefore imperfect and crude,
but, with all its faults, it was entitled to the credit of
being the first publication in which the elevation of all
the continents was ascribed to the weight of the sediment; it was also the first to assert that the elliptical
currents existed before the continents began to rise.
Previous to that time, Sir John Herschel had suggested
that possibly the weight of the sediment, derived from
pre-existing shores, may, in some cases, have raised the
adjacent borders of the continents; but he said nothing
of the causes that raised large areas, and determined
their forms and positions; nor did he propose any theory
of the agency of ocean currents in creating the continents.

Prof. James Hall, of Albany, in his New York
Report, published in 1859, expressed the opinion that
the materials of the Appalachian Mountains were deposited by ocean currents which flowed over that area
while the continent was beneath the surface of the ocean.
He does not say that the mountains were raised in consequence of the depressions thus produced; but after
the deposits were made, and the sunken mass was several thousand feet thick, the whole continent arose,
bringing the mountains up with it. He does not propose any theory, or suggest any cause for the rising of

the continent. In order to account for such a vast quantity of sediment being transported by the currents to this area, he imagines that there was, at no very great distance in the Atlantic, an extensive island or semi-continent from which the sediment was derived.

About twenty years ago, some over-zealous friend of the author accused Prof. Hall of having adopted in his report the ideas published in "Geonomy" two years previous, but the charge was unjust; he proposed no theory to account for the *elevation* of continents or of mountains. The only point on which we agreed was that the ocean currents transported sediment, that by its weight produced depressions. He proposed no theory of the currents, and gave no reason why they flowed in one place or direction rather than another.

In 1866, encouraged and assisted by the Hon. Ira Mayhew, the former Michigan State Superintendent of Education, the author published a new and improved edition of "Geonomy," which, though a pecuniary failure, received the approbation of some of the most distinguished scienticians in our country. In that work the true theory of the elliptical currents was published for the first time, though it was less accurately explained than in these pages.

At the meeting of the American Association for the Advancement of Science, in Chicago, in 1868, the author was elected a member; and on that occasion he read a paper on the ocean currents, in which essentially the same views expressed in this treatise on that subject were advocated. The paper was not only approved by

c

the members present, but Col. Foster (the president-
elect, an experienced geologist) and Prof. Coffin (who
had distinguished himself by his essay on the winds,
published by the Smithsonian Institution) took part in
the discussion, and sanctioned the new ideas advanced.
An abstract of the paper was, without the request of the
author, published in the annual volume of the "Proceedings." It should be stated, however, that he did
not, on that occasion, attempt to prove that the ocean
currents were the agents that, by their operations, created
the continents. He contented himself with getting the
Association to sanction his elliptical theory of the currents; for, this being admitted, the rest of the theory
necessarily follows, as the reader will soon perceive.
Since that time he has published several pamphlets, and
articles in various journals, containing additions and cor-
rections, the results of further investigations.

* The second edition of "Geonomy" excited but little

---

* Lord Bacon, commenting upon the Copernican system, more
than fifty years after the great founder of modern astronomy died,
used the following language:

"In the system of Copernicus there are many grave difficulties.
. . . The introduction of so many immovable bodies into nature,
as where he makes the sun and stars immovable, . . . his making
the moon adhere to the earth, and some other things which he assumes, are proceedings which mark a man who thinks nothing of
introducing fictions of any kind into nature, provided his calcula-
tions turn out well."

Copernicus, when preparing his new system of astronomy,
said,—

"All which things, though they be difficult and almost incon-

interest, and the money that it cost was virtually thrown away. The public generally knew little, and cared less, about geology, and the professional geologists were all apparently absorbed in the practical details of their local surveys. Besides, they had nothing to gain by patronizing a new system which did not originate with themselves, or which did not emanate from some authority higher than their own. As for college professors, their business is to teach established truths, or what seems to be such, and not to introduce novelties. These remarks are made by way of excuse for the long neglect of the author to push his "Geonomy" into public notice. But although it is said that "the mills of the gods grind slow," they move at last with irresistible force. Several advances, which may fairly be termed discoveries, that were first published in "Geonomy," have since been rediscovered by distinguished authors. Reclus, in his "Earth," page 77 (published twenty years after the first edition of "Geonomy"), says, "If we take into the account the geological conditions of Asia, we may, perhaps, be warranted in looking upon the Caspian, the Sea of Azof, and the other lakes of Western Asia, as the remains of the former ocean, which, in the northern hemisphere, formed the equipoise to the Indian seas. There would, then, have been three double oceans, just as there are three continental pairs."

---

ceivable, and contrary to the opinion of the majority, yet, in the sequel, by God's favor, we will make clearer than the sun, at least to those who are not ignorant of mathematics."—*Whewell.*

M. Fay, one of the first scienticians in Europe, director of the Observatory at Paris, in 1881 had the courage and sagacity to advocate the theory first published in "Geonomy" in 1857, that all the continents were elevated in consequence of the weight of the sediment accumulated in the floor of the ocean.

All that is now wanted for these authors to complete their recognition of the system of geonomy is to adopt the elliptical theory of the currents. It is encouraging to find that the new theory begins to be tolerated, and that the idea of the former existence of a North Indian Ocean is no longer regarded as visionary. We may charitably suppose that the authors who have adopted these advances believed them to be original with themselves, but the canons of science imperatively require that the first publishers of a new truth shall be entitled to the credit of priority, especially when several years have passed since the first publication.*

---

* " Dr. Black lectured on latent heat in 1762. Lavoiseur and Laplace borrowed from him, but never mentioned him. Deluc persuaded Black to let him publish the discovery, and then claimed it as his own."—*Encyc. Brit.*

SECTION II.

THE GENERAL OCEAN CURRENTS.

BEFORE commencing the explanation of the new theory of the ocean currents, let us state briefly the facts which are generally admitted, or that will not be denied, and therefore need not be discussed.

1. There are five great oceans, in each of which there is an elliptical current or whirl of the water.

2. None of these ellipses extend farther from the equator than the forty-fifth parallel, and none of them cross the thermal equator. Geographers do not state this fact in words, but they represent it on their maps, though they give no reason for it.

3. In each of the oceans there is a gathering of grass and other materials within the central parts of the ellipse; and this is, with good reason, supposed to be the effect of the elliptical movement of the water.

4. The elliptical currents flow constantly, whatever may be the direction or force of the wind.

5. In each ocean the elliptical current flows due west near the equator, and due east between the fortieth and forty-fifth parallels.

6. The part of an ellipse, or any other current, that

4

flows toward the equator is cold, and the part that flows from it is warm.

7. The warm elliptical current always leaves the vicinity of the equator, in the northern hemisphere, flowing northwest, and in the southern hemisphere southwest. No theoretical explanation of this fact is given by geographers, except that it is supposed to be caused by the wind. A different explanation will be given in another place.

8. All the writers upon the currents assume that they would not flow in ellipses were it not that they are confined within ocean basins, and are deflected from their courses by the shores of the continents.

9. No writer has heretofore recognized the real difference between the causes of the elliptical and the local currents. The elliptical currents have been regarded as mere local currents turned from their normal courses by shores or by winds.

10. The inertia of the currents has only been regarded as influencing them when flowing north or south, but never when flowing due west or due east; they were then supposed to be under the influence of the wind.

11. The elliptical currents are not supposed to abrade the bottom of the ocean, nor, as a general rule, to approach within thirty or forty miles of the shore.

12. It is now admitted by the best authors that the primitive cause of all the constant ocean currents is the difference of temperature in the different latitudes. Halley, one of the most eminent of British philosophers, was the first to assert that the earth's axial rotation is

the cause of the currents flowing in loxodromic directions,—that is to say, when a current of wind or water
is by any cause impelled poleward, rotation deflects
it obliquely eastward also ; and, when it is impelled
equatorward, rotation deflects it westward.

Prof. Huxley, in his Physical Geography, describes
the Gulf Stream as a comparatively superficial current;
and we would add that the water of all the ocean currents together probably constitute, at any one time, not
one-fourth of the whole ocean. The forces that generate
the currents—the cold and heat—act mostly at the surface, while the great cold mass of the ocean waters remain below undisturbed. It is doubtful whether the
cold currents that return from the polar seas reach the
bottom of the ocean where it is very deep.

It is convenient, when discussing the causes of the
currents, to speak of heat and cold, and easting and
westing, as if they were four distinct impelling forces.
But in reality all the ocean currents are produced by
two forces,—gravitation and the inertia of rotation. In
the warm tropical latitudes heat expands and elevates
the water, so that it gravitates and slides down an in-
clined plane toward the pole ; in the high cold latitudes
the water is condensed and sinks, and gravitates toward
the tropics to restore the equilibrium. According to
this statement, heat and cold are the *primary* causes, and
gravitation the *immediate* cause, of all north and south
currents. When this is understood, there is not much
impropriety in speaking of heat and cold as impelling
forces.

The force that impels a current eastward is termed easting, and that which impels it westward is termed westing, and both result from inertia ;* in other words, from the earth's axial rotation.

† The earth moves eastward at the twentieth parallel much faster than it does at the thirtieth or fortieth ; consequently, when a current leaves the twentieth parallel, it carries with it the easterly force which it has acquired there, this force is inertia (*vis viva*). But when a current flows in a contrary direction, instead of carrying a surplus of easterly force, it arrives at its terminus with a much less easterly force and motion than the ocean there possesses. The consequence is that the ocean waters rush

---

* The word inertia *originally* signified the power of remaining passive or inert, but it has now come to be used to signify also the power of continuing in *motion*. Accurate writers, to prevent being misunderstood, sometimes use the term *vis viva*, or living force, to signify force that produces motion, and *vis mortua*, or dead force, to signify force that merely resists motion. The term *vis inertia* is by some used to signify active force, and *inertia* to signify passive force. The force that produces the eastward course of currents is *vis viva*, and that which produces westing is *vis mortua*.

† The rotatory velocity of the globe

		Miles per hour.
At the equator is .	. . . . .	1037
" 5th parallel	. . . . .	1033
" 10th "	. . . . .	1021
" 20th "	. . . . .	975
" 30th "	. . . . .	899
" 45th "	. . . . .	735
" 50th "	. . . . .	668
" 60th "	. . . . .	520

against it and past it eastward, and all that it can do for
a while is to maintain its position and resist the easterly
motion of the ocean waters. This force of passive resistance is also inertia (*vis mortua*). It is termed *westing*,
—not because it really produces a westward movement,
for it does not, but because the things around it move
east, and *leave* it west *of them*, and also because lookers-
on are deceived by appearances, and made to suppose that
it is the current moving west, and not the ocean and earth
moving east.

Every one who has studied astronomy is aware that it
is impossible to form a correct idea of the subject without first becoming acquainted with the laws of the forces
that produce the elliptical orbits. The same is true of
geonomy : it is absolutely necessary, in the very beginning, to clearly understand the elliptical ocean currents,
the forces that produce and that resist them, and the laws
that govern these forces.

The difference of temperature in the different latitudes
is the *primary* cause of all constant ocean currents. All
the currents that flow toward the equator are cold, and
all that flow from it are warm ; all the currents that flow
to the polar seas are warm, and all those that flow from
them are cold.

It is assumed by all writers on the currents, that when
one flows from the tropics it has warmth enough to impel
it to the pole ; and when a cold current flows toward the
tropics, it has coldness enough to impel it to the equator.
It is evident that this is an error, for it is certain that in
each of the oceans the greater part of the warm water

4*

that flows from the tropics does not flow more than half-way from the equator to the pole: it does not flow nearer the pole than the forty-fifth parallel. Now, if it is the warmth, and nothing else, that impels a current toward the pole, it follows that when the current ceases moving poleward, at the forty-fifth parallel, it is because its warmth is exhausted, or is insufficient to impel it farther. So, also, if it is coldness that impels a current toward the equator, and if it ceases moving in that direction before it quite reaches the thermal equator, as the elliptical currents all do, the reason is that it has lost its impelling cause—its coldness. Again, if a large current flows from the tropics, and when it is moving poleward it divides into two branches, and one branch flows on to the polar seas, while the other ceases flowing in that direction at the fortieth or forty-fifth parallel, we know that the reason is that the branch that flows to the polar seas is the warmer, and the branch that falls short is the colder. These facts, which appear to be very simple, are important, as we shall hereafter perceive, though they have hitherto been overlooked.

* Another great mistake has been made, even by writers

---

* "Sir Charles Lyell, referring to the effect of the rotation of the earth on its axis upon the currents, says it ' can only come into play when the waters have already been set in motion, and when the direction of the current happens to be from south to north or from north to south.' "—*Prin. of Geol.*

" The earth's rotation is not supposed to be a cause of motion in the waters; but there being a movement for other reasons, it gives (in the northern hemisphere) easting to the flow in the

of reputation ; they have assumed that the earth's rotation only affects the currents while they are moving north
or south. In the northern hemisphere, they state cor-
rectly that a current which is impelled due north is
deflected eastward by rotation, so that it really moves
northeast; but when it reaches its northern terminus
they assume that the easterly force, derived from rotation,
ceases to operate. This is an error ; they have strangely
overlooked the real nature and effects of inertia.

The truth is that when a current from the tropic
reaches the forty-fifth parallel, although its northerly
force—its extra warmth—may be exhausted, its easterly
force—its *vis inertia*—is not: it continues to act alone,
and impels the current due east. So also, when a cold
current flows within five or ten degrees of the equator,
and loses its extra coldness, and can flow no farther in that
direction, its inertia acts alone, and impels it relatively
due west.

The earth, in the twentieth parallel, at the Gulf of
Mexico, moves eastward more than two hundred miles
per hour faster than it does at the Banks of Newfoundland, in the forty-fifth parallel. When the water from
the Gulf reaches the Banks, though its extra warmth is
nearly or quite exhausted, so that it can move no farther
north, it still retains the greater part of the easterly force
which it brought with it from the Gulf, and which impels it due east nearly across the Atlantic. This con-

northern direction, and westing to the flow in the eastern
direction."—*Dana.*

servation and persistence (*vis viva*) of the easterly force
has been ignored, or under-estimated, by geographers.
They have been familiar with the *fact* that the current
flows due east from the Banks* of Newfoundland more
than two thousand miles, and then turns south, and they
have imagined that it was turned out of its proper northern course by the wind, by the Banks, or by the cold
current that flows from the Arctic Sea. But when we
consider that the current leaves the Gulf with an easterly motion of about two hundred miles per hour greater
than the ocean at the Banks possesses, and carries nearly
all this extra easting to the Banks, we perceive that there
is no need of invoking the wind, or any other additional
or imaginary force, to impel it due east when its warmth
is exhausted. If a man could be instantly transferred
from the middle of the Gulf to the Banks, he would find
himself moving eastward at the rate of two hundred miles
per hour; or if he could, on the contrary, be transferred
from the Banks to the Gulf, he would find the earth
there moving past him eastward and the west coming
toward him at the rate of two hundred miles per hour.

When these facts are fully appreciated we shall cease
speaking of the easterly force as merely an attendant of
the north and south currents. . . . The force derived
from rotation not only acts in some places separately, and

---

* "The current passes the Newfoundland Bank and stretches
over toward Europe, then a part bends southeastward to join the
tropical current and complete the ellipse, the centre of which is
the Sargasso Sea."—*Dana.*

independently of the north and south forces, but it is a
much greater force than they. This is demonstrated by
the fact that the diameters of the ellipses, in all the
oceans, are greater east and west than north and south.

We often speak of the trade-wind blowing west, and
of the equatorial current flowing west. It is often convenient, and perhaps generally harmless, to use this
language, but in reality neither the wind nor the water
moves west at all. Let us explain. When a cold current from the north approaches within five degrees of the
equator, and loses its coldness, so that it can flow no
farther south, its inertia (lack of easterly force) enables
the western part of the ocean to move eastward past it,
and thus bring the current into a relatively more western
position. The process may be familiarly illustrated by
laying a sheet of paper on a table, and with the left
hand moving it slowly eastward, while, with the right
hand, a pencil is moved due south on the paper. The
line drawn will be a curve southwest. When the pencil
is near the top of the sheet of paper (which may represent the equator), if it is moved eastward very slowly,
while the paper is moving east much faster, the pencil-
mark will be drawn due west, though the pencil has not
moved west at all, but, on the contrary, has slowly
moved east.

When the current flows from the Gulf to the Banks,
it really does move eastward faster than the earth and
ocean beneath it, and it continues to do so after it reaches
the Banks, and ceases to move north. . . . To illustrate
this with our sheet of paper, we may move it east as

before with the left hand, and with the right hand move the pencil east *faster* than we do the paper with the left.

These illustrations enable us to perceive that it is un-necessary to attribute the apparently westward move-ment of the equatorial current to the winds, as so many writers have done. The trade-winds themselves do not blow west, though they seem to do so. They come from the higher and colder latitudes, as the ocean currents also do, and bring their inertia (lack of easterly force) with them. It would be quite as correct to say that the ocean currents cause the trade-winds as that those winds cause the equatorial currents.

The writers upon this subject seem to have assumed that the force communicated by rotation is all expended as fast as it is received, so that when, in the northern hemisphere, a current reaches its northern terminus, it is destitute of easterly force; and when it reaches its southern terminus, it is destitute of westerly force. When, therefore, they are called upon to account for the due easterly or due westerly courses of the currents, they have recourse to the wind. The idea that the easterly and westerly currents are influenced by inertia alone does not seem to have occurred to any of them.

It is undoubtedly true that the trade-wind, at the equator, apparently moves west faster than the equator-ial current, since it fills the sails of vessels and drives them westward more rapidly than the current moves; and this is probably the reason why the trade-wind has been supposed to produce the current. It is worth while, therefore, to inquire why the wind appears to

move west faster than the current.* The most reasonable explanation appears to be, that the wind moves
more rapidly north and south than the ocean current
does, and therefore loses less of its easterly or its westerly force (its inertia) on the way. When, for example,
the wind leaves the thirtieth parallel, it has the slow
easterly motion proper to that latitude, and when it
reaches within five degrees of the equator, it is still
nearly as slow as when it started. The slower the *easterly* motion of the trade-wind really *is*, the more rapid
is its *apparent westward* movement; for, in reality,
nothing near the equator moves westward,—everything
is moving eastward; but the earth itself, including the

---

* Sir Charles Lyell, referring to the opinion that the deep currents are caused by the winds, quotes the observations of Darwin,
that "notwithstanding the great force of the waves on the South
American shore, all rocks sixty feet under water are covered
with sea-weed,—the effects of the wind being only comparatively
superficial."—*Prin. of Geol.*

Prof. Guyot, the most accurate of geographers, says, "The same
forces drive both the wind and the ocean currents in a common
direction.

"A knowledge of the one will facilitate the understanding of
the other.

"The equatorial current is analogous to the trade-winds, it has
even been thought that these winds were the cause of it; but it is
too deep and rapid to admit of being explained by their action
alone.

"Since the general winds, as we have seen, owe their origin to
this same cause, we shall not be surprised to find a similarity, and
in some cases a remarkable coincidence, between the march of the
atmospheric currents and that of the currents of the ocean."

general ocean and atmosphere, moves eastward fastest, the equatorial current next fastest, and the trade-wind the slowest, and for this very reason it seems to move west the fastest. If two men run a race east, the one who runs the slower will get continually farther and farther west of his opponent, though he is all the while running east. It is only in this sense that winds, or ocean currents, may be said to move west. Suppose a current of wind and another of water to start together from the Gulf of Mexico (twentieth parallel), and move to the Banks of Newfoundland (forty-fifth parallel), and suppose that when they arrive there they both cease to be warm, and begin to move east,—would they move with equal rapidity? probably not. In the first place, the wind will arrive at the forty-fifth parallel much the sooner, and bring more of its easterly force (inertia) with it; the water current will arrive long afterwards, and bring less easterly force, and both currents will move due east, but the wind the faster; now, although we should not say that the wind *produced* the water current, we might, with some reason, say that it increased it, at least at the surface.

The winds blow relatively west near the equator, and east near the forty-fifth parallel, in all the oceans; they cannot do this without moving in ellipses. If they move in ellipses now, they surely did when there was no land to obstruct or deflect them. If, therefore, any one asserts that the winds *cause* the ocean currents to flow in ellipses, the statement, even if true, is no serious objection to our geonomic theory. We contend that the ocean currents moved in ellipses of a certain form when there was no land,—the cause of that movement, though important, is a secondary question.

The probability is that the reason why scienticians have neglected to investigate the laws of the currents thoroughly, and to discover the truth concerning them, is, that they have not regarded them as of much importance. Had they suspected that the currents, by their operations, created the continents, they would long since have wrung from them all their secrets.

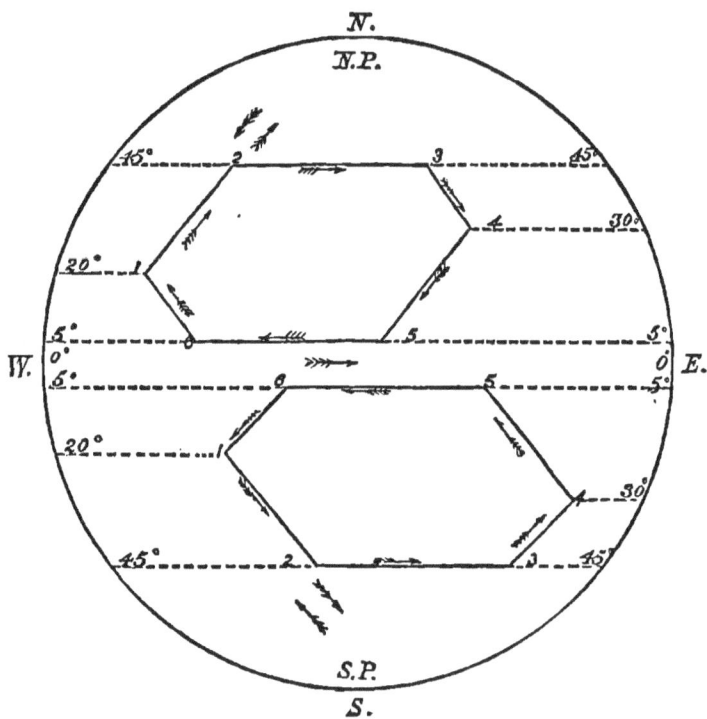

FIG. 2.—A PAIR OF ELLIPSES.

Fig. 2.

This figure perfectly illustrates the elliptical theory. The two ellipses are limited to the two zones between the fifth and the forty-fifth parallels. Between the ellipses is an interval, in which an arrow indicates the course of the equatorial counter-current.

Between the pole and the forty-fifth parallel, in both hemispheres, is a large area in which the Polar Seas became land-locked during the Glacial Epochs. The two arrows north and the two south of the forty-fifth parallels represent the currents to and from each ocean, and the Polar Seas, the exclusion of which produced the Glacial Epochs.

The current in the northern hemisphere starts warm from the twentieth parallel (1), and flows northeast to the forty-fifth (2), it then, being neutral in warmth, flows under the influence of inertia due east to (3), and then, becoming cold, flows southeast to the thirtieth parallel (4), where, its easting being exhausted, its inertia causes it to flow southwest to the fifth parallel (5); here, its extra coldness being exhausted, its inertia causes it to flow due west to (6), when its warmth causes it to turn northwest and flow to the twentieth (1). The ellipse in the southern hemisphere is the same except that the current flows south from the equator and north toward it.

The ellipses in this figure and also in Fig. 1 are made angular, to indicate the turning points more precisely, but in reality the currents turn in curves.

SECTION III.

THE ELLIPTICAL CURRENTS.

THE nature of the forces that operate to produce all the ellipses* may be sufficiently illustrated by the one that circulates around the North Atlantic basin. This current starts warm from the Gulf of Mexico and its vicinity, near the twentieth parallel, and flows northeast to the Grand Banks of Newfoundland. There it divides; one branch, which is still warm, continues to flow northeast to the British, the Baltic, and the Arctic Seas; but the main branch, having lost its extra warmth and retained its easting, flows due east from the Banks nearly across the ocean; but before it reaches Europe, its increasing coldness causes it to turn and flow southeast to the thirtieth parallel; at this point its easterly force is exhausted. The current has now performed just half of its elliptical journey around the ocean. It has

---

* Referring to the fact that in each ocean there is an ellipse, Prof. Dana, in his Manual, remarks, "The plan or system for each ocean north and south of the equator is the same. A flow in either tropic from the east, and in the higher temperate latitudes from the west, the one flowing into the other, making an elliptical movement."

moved northeast, due east, and southeast. In performing the other half it will move (apparently) westward instead of eastward (that is to say, it will move southwest, due west, and northwest). It starts cold from the thirtieth parallel, with only the easterly force proper to that latitude, and flows southwest nearly to the equator before the coldness, which impels it* south, is exhausted. But its inertia—its lack of easterly force—remains and impels it—apparently—due west nearly across the ocean. Before it reaches the South American coast it becomes extra warm, and, therefore, turns northwest, and continues in that direction to the twentieth parallel, in the Gulf of Mexico, from whence it started, and thus it completes its elliptical circuit.

If we make a drawing of this ellipse, we observe that it consists of two halves, in one of which the current moves more or less easterly, and in the other westerly. In moving through the first half there is an excess of easting, and in the second half there is a lack of easting,—inertia,—the practical effect of which is the same.

The easterly force is exhausted at the thirtieth parallel on the eastern side of the ocean, and the westerly force

---

* "Of the mass of the water that is brought into the Mid-Atlantic by the Gulf Stream, it may be stated, with confidence, that the larger proportion turns southward to the east of the Azores, and helps to form the North African current.

"The Gulf Stream when last recognized as a current is flowing due east, and its southern portion turns first southeast and then south."—*Ency. Brit.*

—inertia—is exhausted on the western side at the twentieth parallel.

In this treatise it is assumed that when the ocean covered the globe the elliptical currents, in the northern hemisphere, flowed west near the equator; that they turned northeast at the twentieth parallel, due east at the forty-fifth, and southwest at the thirtieth; but it must not be supposed that these were then, or are now, *precisely* the turning-points. At the present time there are several causes of variance, some of which did not exist then. Among the disturbing causes are :

1. The unequal distribution of land in the two hemispheres.

2. The great elevation of land within the Antarctic circle.

3. The land-locked condition of the Arabian and Bengal Seas.

4. The communication of the Atlantic with the Arctic Sea.

5. The great extent of land that has resulted from the elevation of the floor of the North Indian Ocean.

6. The narrowing of the oceans and extension of the continents.

In the Pacific the thermal equator and the westward equatorial current are, even in July and August, on the north side of the geographical equator, and in January they are not less than ten degrees north of it. In the Indian Ocean, on the contrary, the equatorial current in July and August is between ten and twenty degrees south of the equator; consequently, the whole ellipse is carried south. A large part of the current in this ocean flows as far south as the Cape of Good Hope before it turns east. The difference between the currents of these two oceans can only be understood after considering the difference in the positions of the lands north of them. The North Pacific is almost entirely cut off from the Arctic Sea. Behring's Strait is only thirty miles wide, and so shallow that but little cold arctic water can flow through. This ocean is less cold than any other, and the current, therefore, has less extra coldness to lose before it is in a condition to flow west. But the South Pacific is larger and colder, and its cold current must flow farther north to acquire the thermal condition necessary to qualify it to flow due west.

In the Indian Ocean the land-locked Arabian and Bengal* Seas, north of the equator, are the warmest in the world, and, consequently, the current from the south loses its extra coldness and flows due west ten to twenty degrees from the equator. When the causes of the elliptical circulation are clearly understood, it is easy to prove that originally the ellipses were necessarily alike; but when we apply the same principles to the currents as they flow at the present time, and endeavor to show why they differ from each other, we need correct information con-

---

* The monsoon, or season-winds, of the Arabian and Bengal Seas are supposed to cause analogous changes in the currents of the Indian Ocean ; but it is much more reasonable to suppose that the same causes that change the winds change the courses of the currents also ; the writers upon this subject seem to have mistaken coincidences for causes.

cerning the peculiar circumstances of each ocean which
tend to vary its elliptical circulation.

But notwithstanding all the existing causes of variance, the ellipses, the oceans, and the continents vary from
their primitive forms too little to render the elliptical
theory of the origin of the continents in the least degree
doubtful. Perhaps the principal reason is that the plan
of the oceans and continents was laid out, and their
forms and positions fixed, before they began to emerge
from the ocean, and were subjected to the action of the
waves,—the unequal distribution of sediment from the
shores,—and the invasion of floods and glaciers from the
circumpolar regions.

SECTION IV.

EFFECTS OF OCEANIC RESISTANCE UPON THE FORM OF THE ELLIPSE.

It will be noticed that the ellipses are represented as of an irregular rhomboidal form, instead of being symmetrical and oval. As this peculiarity has never before been referred to by any writer, it is proper to give it special attention. When the current starts from the twentieth parallel, on the western side of the ocean, it has the easterly force proper to that latitude; and if it could move through the ocean (as planets are supposed to move through space) without encountering any resistance, or losing any of its force on the way, it would continue to move easterly (northeast, due east, and southeast) until it reached the twentieth parallel again on the opposite (eastern) side of the ocean. But it does encounter resistance; and it expends so much force on the way to overcome it that it falls short about ten degrees, and ceases moving southeast at the thirtieth parallel. In performing the second half of its elliptical journey it starts from the thirtieth parallel and moves southwest, due west, and northwest, but only reaches the twentieth parallel on the opposite (the western) side

of the ocean before it ceases moving west. It is necessary to understand this irregular form of the ellipses,
and the causes of it, since it is impressed upon the outlines of the continents, and thus confirms the elliptical
theory of their origin. If the outlines of the continents
are dependent upon those of the ellipses, there must of
course be a general coincidence between them.

It should be stated that the peculiar form of the ellipses was not discovered by observation, but by mathematical reasoning. Tell any mathematical engineer that a
current starts at the twentieth parallel, on the western
side of the ocean, with the easterly motion of that latitude, and that it is expected to move through the ocean
to the same latitude on the eastern side, and he will, without hesitation, assert that the resistance of the inert ocean
will cause the expenditure of a certain quantity of the
easterly force, so that it will fall short of the twentieth
parallel, and begin to move southwest before it reaches
there. How much it will fall short will depend upon
the degree of resistance, and this can only be ascertained
by observation.

When the current leaves the thirtieth parallel on the
eastern side of the ocean it has a certain amount of
easterly force,—no more and no. less than the earth has
in that latitude. It has less easterly force than the ocean
waters through which it now passes, but it has inertia,
and if this could remain unchanged (if the current could
receive no more easterly force on the way) it would continue to move (relatively) west (southwest, due west, and
northwest) until it reached the thirtieth parallel again

on the western side of the ocean. But it does continually receive accessions of easterly force (loses more and more of its inertia), and consequently it falls short, and only reaches the twentieth parallel before it ceases moving west. The reader will perceive that this falling short varies the form of the ellipse. Another circumstance that prevents the ellipse from being perfectly oval is that the inertia of the currents causes them to flow *due* west near the equator, and due east* near the forty-fifth parallel, consequently, the east and west sides of the ellipses are, for a short distance, straight instead of being curved.

---

* "In its course to the north the Gulf Stream gradually trends more and more to the eastward until it arrives off the Banks of Newfoundland, where its course becomes due east."—*Maury.*

SECTION V.

LACK OF SYMMETRY.

IF we look at any map of the world we may observe
that each of the southern oceans is about fifty degrees
of longitude east of its northern mate: the South Atlantic is east of the North Atlantic; the South Pacific
is east of the North Pacific; and the South Indian is
east of the area that was formerly occupied by the North
Indian. The positions of the continents are equally unsymmetrical: South America is east of North America;
South Africa is east of Europe; and Australia is east of
Asia.

The obvious explanation is that the elliptical currents commenced their career before the ocean basins
were formed or the continents began to rise. As each
ellipse was independent of every other, there was no
physical necessity for their being arranged symmetrically north and south; indeed, the chances were against
such an arrangement. If we place a series of oval
bodies, eggs for example, in a row, and then place
another row parallel and in contact with them, and
arrange them symmetrically, a little jolting will derange their symmetry, and probably cause them to

assume nearly the same relative positions as those that the ellipses and oceans now actually occupy. Besides, if one southern ellipse were from any cause placed fifty degrees east, the other two would be forced to follow. Whatever may have been the cause, the fact is certain that all the oceans and continents north of the equator are placed west of their southern mates.

The principal deranging effect of this lack of symmetry is upon the forms of the central, or tropical parts of the continents. This effect may be well illustrated if we draw an ideal map of the world, in which the oceans and continents are placed symmetrically north and south, and then run a knife through the map at the equator, and push the southern half a short distance east of the northern, and observe the effect upon the central continents. The two halves are drawn apart, and made longer and narrower, so as to resemble Central America; and, like that part of the continent, the trend will be loxodromic. The northern oceans also are pushed relatively westward, so as to form a gulf on the *eastern* side of the northern continent. The Gulf of Mexico was in part produced in this manner; so, also, was the China Sea. The western prolongation of Northern Africa was formerly a gulf, analogous to that of Mexico, and produced in the same manner.

The southern oceans are similarly pushed eastward into the *western* sides of the southern continents. This is well represented and strongly marked by the Gulf of Guinea, and in a lesser degree by the western side of Panama.

Geographers have often remarked the resemblance of
the position and trend of Central America to the chain
of islands connecting Asia with Australia. The explanation is found in the fact that both were produced by the
original unsymmetrical arrangement of the ellipses before
the continents began to rise. The western border of
Northwestern Africa was at first analogous to Central
America.

SECTION VI.

LOCAL AND COUNTER-CURRENTS.

WHEN any part of the ocean becomes extra warm or extra cold, a current is generated, and the water, if warm, flows toward the pole, and if cold, toward the equator. A local current in the northern hemisphere, if warm and unobstructed, always flows northeast, and if cold, southwest. In the southern hemisphere, if warm, it flows southeast, and if cold, northwest. An elliptical current flows in all directions, and it sometimes happens that, in a part of its course, a local current flows parallel with it, but in a contrary direction. These local counter-currents have greatly puzzled geographers. As they made no distinction between the causes of local and of elliptical currents, and understood very little concerning the effects of inertia in forcing currents east and west, they were unable to understand that the counter-currents are regular and normal but local currents.

There is a remarkable counter-current in the North Pacific,* which Prof. Page and Alexander Keith John-

---

* "A third part of the excess of water drawn westward escapes back again eastward between the main branches of the equatorial

son and other geographers describe as "an equatorial counter-current."

It flows eastward, according to Johnson's map, in the interval between the two great westward equatorial currents, nearly the whole length of the ocean, from China to America. He *asserts* that it is the *same* westward equatorial current that, having reached its western terminus, has turned back, and escapes eastward. These highly respectable authors ascribe the great equatorial currents to the trade-winds. Is it not strange that it has not occurred to them that this large counter-current flows eastward in the very face of the same trade-wind that is *supposed* to blow the other two great equatorial currents westward?

The fact is that the water in the space between the two great equatorial currents is warm, and of course it generates a series of local currents throughout its whole length, which would flow northeast if they were not, so to speak, fenced in, and forced to flow due east between the two great currents that are flowing westward.

Lieut. Maury, in his "Physical Geography of the Sea," states the fact that the trade-wind commences in

streams, and forms the counter-current known as the Guinea current." "In the Pacific part also returns toward America as a counter equatorial stream."—*Phys. Geog.*, by David Page, F.R.S., etc., Edinburgh.

According to this author the trade-wind blows two-thirds of the water near the equator due west, and at the same time tho other third "*escapes back*" due east in spite of the wind.

about the thirtieth parallel, in the northern hemisphere, to blow southwest, and in the same latitude in the southern hemisphere to blow northwest. But he frankly confesses that he can find no reason for it. We have given the reason why the ocean currents commence flowing southwest in the northern hemisphere and northwest in the southern in these latitudes, and of course the same explanation applies to the winds. They are governed by the same general laws as the ocean currents, but are varied much more by local circumstances.

Geographers have hitherto been unable to give a reason for the fact that in the northern hemisphere the current flows southeast from the fortieth or forty-fifth parallel to the thirtieth and then turns and flows southwest; nor have they understood why the warm current flows from the equator northwest to the twentieth parallel and then turns and flows northeast.

Those authors who attribute wholly to the wind the current that flows due east from the Banks of Newfoundland, should give some reason why the same wind does not impel *the whole* of the Gulf Stream due east, instead of allowing it to divide, as it does, at the Banks,—one part flowing due east and then southeast, and the other part flowing *northeast* to the Arctic Sea. They should also explain why it is that, in the face of the same wind, a return current flows southwest *from* the Arctic.

The Guinea current in the Atlantic is another local warm equatorial counter-current, generated in the interval, between the two great westward currents, near the equator, and flowing southeast into the Gulf of Guinea, while the great elliptical current is flowing in a contrary direction. The moment the distinction is clearly understood between the causes of the local and of the elliptical currents, the mystery of the counter-currents disappears.

The Antarctic Drift Currents.—In the southern oceans, between the Antarctic coast and the forty-fifth parallel, there is apparently a general drifting of the surface of the ocean eastward. For want of a better reason, the geographers, including the distinguished mathematician Mr. Croll, ascribe this easterly drift to the winds that blow in the same direction there. Mr. Croll admits that these currents from the Antarctic would naturally flow northwest were it not for the great force of the wind that blows east. He does not pretend that the wind comes from the Antarctic coast, for if it did, he well knows that it would not blow east, but northwest. He must admit that it comes from the tropical region, and *for that reason* it blows east. Surely the same reason applies to the ocean current. It comes warm from the tropical latitudes, and when it reaches the forty-fifth or fiftieth parallel it is cold, and can flow no farther south; but its easting is not exhausted, and consequently it flows due east. It is perfectly analogous to the currents that flow northeast and due east from the Grand Banks of Newfoundland. There is only one of the northern oceans that has free access to the Arctic Sea, but there are three great oceans in the southern hemisphere, each of which send offsets into the cold Antarctic seas, and these offsets unite to create an apparent general "drift" of the surface of the ocean eastward.

Prof. Houston, of Philadelphia, and several other respectable geographers, represent the currents as approaching the equator from the north and from the south, meeting there in antagonism, and moving westward together. But the warm counter-current that flows east in the equatorial interval, proves that the great currents lose their northern and southern forces before they reach the thermal equator, for otherwise there would be no such interval and no such eastward equatorial counter-current.

It is now asserted by the highest authorities, including Dr. Carpenter, that the movements of the currents north and south are caused by differences of temperature. This being admitted, when a warm current moves north to the forty-fifth parallel, and then ceases to move in that direction, it follows that it is *because* its extra warmth is expended. So, also, when a cold current moves to within five or ten degrees of the equator, and then ceases to move in that direction, it follows that it is because it has acquired the warmth normal to that latitude.

SECTION VII.

LIMITS OF THE ELLIPSES.

THERE were originally just three ellipses in each hemisphere, and all were within the zone between the equator and the forty-fifth parallel. If the warmth of the currents from the tropics had been much greater they would all have continued flowing poleward far beyond the forty-fifth parallel. The fact that none of them approached nearer to the pole is proof that they lacked the requisite thermal force. When the water leaves the tropics it is not all equally warm; and, on its way, those parts that are brought into contact with the air or the cold ocean water lose their extra heat and cease moving poleward, while the warmer parts continue flowing toward the pole. We know by observation that in all the oceans the greater part of the current ceases moving poleward before passing beyond the forty-fifth parallel, and we know that the current of warmer water continues to the Polar Sea.

When the ocean covered the entire globe there must have been an almost perfect equality between the ellipses: they probably all extended poleward to near the forty-fifth parallel; and as the season changed they swerved north and south across the geographical equator.

We have seen good reasons why the ellipses are all limited north and south. What is it that limits them east and west, and causes them to form just three ellipses and no more in each hemisphere?

When the elliptical current reaches the thirtieth parallel its easterly force is exhausted, it cannot therefore flow any farther east; so, also, when it reaches the twentieth parallel, it can move no farther west, for its inertia (westerly force) is exhausted there. The forces that produce the ellipses are limited in all directions; they can only impel a current north to one point and south to another,—east to one point and west to another, as we have already explained; they could, therefore, originally produce only a definite number of ellipses in one hemisphere, and the same number in the other. To a mathematical and philosophical mind, the fact that there were just three oceans north of the equator and three south of it is very suggestive; the fact that each of the ellipses is limited to the zone between the equator and the forty-fifth parallel is still more so; when to this we add the fact that all the remarkable analogies of the three pairs of continents are confined to the same zones, the inference that the relation is one of cause and effect becomes irresistible. We shall demonstrate in another place that the limitation of the ellipses was the real cause of those hitherto mysterious events that occurred in the circumpolar regions during the so-called glacial epoch.

Judging by the geological indications, and the positions of the land around the Arctic Sea, the elliptical currents must have been nearly, if not quite, the same in extent when the lower Laurentian rocks were deposited as they are now. Would this have been the case if the heat of the sun and the internal heat of the earth had been much greater at that time (millions of years ago) than at present?

That the solar system, including the earth, has been condensed from extremely attenuated matter there can scarcely remain a doubt; but it does not follow that the sun or planets were ever hotter than they are now, or that the condensation resulted from cooling and shrinking. It is a much more reasonable theory that the condensation was produced by pressure, collisions, and chemical changes combined. Oxygen and fuel may both be cold as ice, yet, when combined, they produce condensation, accompanied with radiations of heat and light. Geonomy furnishes no evidence that the earth was ever hotter than it is now, or that its axis or its centre of gravity have ever changed. The indications are that all the changes of the earth have resulted from " causes now in operation."

SECTION VIII.

EXTENSION OF THE AREAS OF THE CONTINENTS AND CONTRACTION OF THE BOUNDARIES OF THE OCEANS.

WE all know by observation that the elliptical whirl of wind or of water tends to collect in the centre of the whirl the floating and the sedimentary materials that come within the sphere of its operations. The sargasso seas, as the collections of grass in the oceans are called, not only illustrate this tendency, but they inform us, in language plainer than that of words, that the same centralizing process must have begun when the ocean was created, and that it has continued without interruption to the present day. Reason also informs us, without the need of observation, that if the hundredth part of an inch of organic or meteoric matter were thus accumulated on an ocean's floor in one year, or one century, after a definite number of centuries (a much smaller number than, according to geology, the oceans have existed) the pile would nearly reach the surface,—unless its weight caused the crust to sink and create an ocean basin.

Furthermore it is certain that if six such sinking basins existed, three each side of the equator, between the equator and the forty-fifth parallel, the fluid or plastic material below the crust would be forced into the inter-elliptical spaces and elevate them, thus giving birth to three pairs of continents. Furthermore, as the basins continued to sink and the continents to rise, the continental areas would become wider, and the ocean basins narrower, for the reason that the sinking would be greatest in the centre. It is also evident that such changes would render it necessary for the crust to stretch or break in order to accommodate itself to the increased areas of surface which the depressions and elevations would produce. Now, what are the facts furnished by geology? The continents, since the first lands emerged, have continued to become wider and the surface of the ocean to the same extent narrower. From the Laurentian age to the Neopliocene period, the lands continually advanced and forced the oceans to retire. The geological history of every continent illustrates this truth.

The original tendency of the ellipses was to make the six ocean basins equal in extent; but when we look upon a map of the world we find the Pacific very large, and the Atlantic, particularly the North Atlantic, small. The ocean currents themselves furnish no reason for this difference. The cause must be found in the unequal manner in which the sediment was distributed. We have no means of knowing the causes of the unequal supply or distribution of the sediment before the lands emerged, and "we can only reason from what we know." If from any cause one ocean received a much greater quantity of sediment than the others, its central area would sink more, and consequently its edges would rise and emerge nearer to the centre, and thus make the basin narrower. Applying these remarks to the North Atlantic, geology supplies us with evidence that it was formerly wider than it is now. The Gulf Stream—a section of the elliptical current—flowed northwestward in the twentieth parallel, over the continent, beyond the Rocky Mountain area, and curved around and flowed east near the forty-fifth parallel.

The sinking of the floor of the Atlantic raised a long low Appalachian island, on which the coal plants grew before the present mountains were created. At the south end of this island the Gulf of Mexico was afterwards located, and at the north end the Gulf of St. Lawrence. The elliptical current flowed west and northwest and east around this island,—entering by the area of the Gulf of Mexico, and escaping by the area of the Gulf of St. Lawrence. When the bottom of the northern part of the interior American sea became elevated, so that the current could not pass through, it was forced to turn back on itself and escape southward,* and then to flow northward on the eastern side of the Appalachian land, which was then a peninsula.

---

* "The absence of sediments from a large part of the continental region must have been owing to the absence of the conditions on which their distribution depends. The currents of the ocean which ordinarily swept over the land (the Labrador current from the north and the Gulf Stream from the south over the interior) must have had their action partly suspended. This may have been caused by a barrier," etc.—*Dana's Manual on the Lower Silurian Age.*

As the land continued to advance, encroaching upon the interior American sea, the waters retreated southward until all that is left of that sea is the present Gulf of Mexico. But even there the contest between the land and water continues; the land is encroaching upon the Gulf, not only by the imperceptible rising of the continent, but also by the addition of the vast quantities of sediment furnished by the Mississippi, and by the labors of an infinite number of coral-building animalculæ.

The former condition of South America was analogous to that of North America. The warm elliptical current from the tropics probably flowed southwest and passed into the interior of the continent near the area now occupied by the mouth of the Amazon, and then flowed south and southeast and passed out through the area now occupied by the mouth of the La Plata. The area occupied by the Brazilian mountains* was then an island analogous to the Appalachian island. When the interior parts of the continent arose the elliptical current was excluded and forced to flow as it does now, on the eastern instead of the western side of the mountains.

---

* "In Brazil there are four parallel ranges of mountains from the plains of La Plata on the south to those of the Amazon on the north, and spreading inland for nearly eighteen hundred miles in a broad plateau, whose mean elevation is three thousand two hundred feet. These ranges or ridges of the table-land are separated from each other by the affluents of the Amazon and the St. Francisco on the one hand, and by those of the Paraguay and Parana on the other, and succeed each other, ridge and plain, with wonderful continuity; the Vertentes is the last, and gradually descends to the great plain of the continent."—*Page's Phys. Geog.*

The great interior North American basin was once analogous to the present Caribbean Sea, and the Appalachians were analogous to the Antilles. The elliptical current now passes through the Caribbean Sea and the Gulf of Mexico, and makes its exit through the Florida Channel. Let that channel be obstructed or blocked up by the elevation of the bed of the sea, and the current would be forced to turn back on itself southward, and escape on the eastern side of Cuba. If, then, the beds of the Caribbean and Mexican Seas should become dry land, the Antilles would be mountains analogous to the Appalachians, and the sea-bed analogous to the Mississippi and St. Lawrence Valleys.

The North and South Atlantic ellipses are doubtless shortened east and west by the narrowness of the ocean basin; this is, perhaps, one of the reasons why the Gulf Stream presses westward against the eastern side of the continent of America, and why a part of it pushes its way so far west in the Gulf of Mexico. The Atlantic Ocean appears to be too narrow for the natural capacities of the ellipse, and perhaps this is the reason of the great rapidity of the Gulf Stream. If it should become yet narrower, the ellipse might still exist, though " cribbed, cabined, and confined." The elliptical current which formerly existed in the North Indian was shortened and restricted more and more until it finally ceased to circulate. If the cooling and shrinking theory were true, would not all the ocean basins have become gradually wider and the continents narrower?

SECTION IX.

LOXODROMIC TRENDS.

THE trends of the principal shores of the continents, instead of being directly north and south, or east and west, are loxodromic,—that is to say, they are either northeast or northwest.* No one has hitherto been able to suggest a reason for this fact. Humboldt supposed that the cause was hidden below the crust of the globe, and consequently that it would never be known. The moment, however, that the light of the elliptical theory is brought to bear upon the subject the mystery disappears. All the north and south ocean currents, as we have seen, necessarily trend in loxodromic directions, and, of course, the shores of the continents which they created coincide with them. If it is objected that the currents coincide with the shores because they are de-

---

* " Two great systems of courses or trends prevail over the world, a northwestern and a northeastern, transverse to each other. The islands of the oceans, the outlines and reliefs of the continents, and the ocean basins themselves alike exemplify these systems.

" While there are many variations in the courses of the earth's feature lines, there are two directions of prevalent trend,—the northeasterly and the northwesterly."—*Dana, Manual,* p. 29.

flected by them, and that if the shores were not there
the currents would flow in other directions,—a perfect
answer is found in the theory of the currents as taught
by all geographers. It has been demonstrated that they
would move in the same loxodromic directions if the
shores did not exist.

If lands near the equator should emerge from the
ocean their trends would not be loxodromic, but east
and west, for the reason that the currents there all flow
in those directions. Why have not lands risen near the
equator?

1. There are two equatorial currents there that flow
due west.

2. Between these two there is in the Atlantic and
also in the Pacific a powerful counter-current flowing
east. These three currents distribute sediment and produce depressions so as to prevent elevations at the
equator.

3. The changes of the seasons cause these three currents to swerve north and south over a broad zone.
Even in the southern parts of the North Indian Ocean
there is a broad area north of the equator, including the
Arabian and Bengal Seas, in which there is no land,
for the reason that the equatorial currents once flowed
there.

The northern shore of the North Atlantic would undoubtedly have trended east and west were it not for the
powerful north and south currents that flow to and from
the Arctic Sea. Even as it is there are some reasons for
the opinion that the so-called submarine "telegraphic

plateau" is only a submerged land, with an east and west shore.

The north shore of the North Pacific has in the main an east and west trend, and thus coincides with the course of the elliptical current in that latitude.

SECTION X.

SEDIMENT.

It cannot be proved directly, by ocular demonstration, that the weight of sediment has produced a single depression of the earth's crust; but we have indirect and circumstantial evidence of the most satisfactory character. When we see the impression of a human foot in the sand on the sea-shore, although we did not see it made, we know its origin by its numerous coincidences with real human feet. It is useless to object that a combination of accidental circumstances might have produced the resemblance: no one will believe it. When in addition we see six successive impressions of the feet, at such intervals and in such relative positions as generally occur between men's footsteps, all doubts vanish. So, also, when we find on one side of the equatorial line three ocean basins, and on the other side three similar basins, in each of them an elliptical current, and all the ellipses limited to a particular zone; when we find three pairs of continents very analogous to each other, each continent located between two ellipses, with all the analogies of the continents limited to the same zone, and the outlines coinciding with the lines of the ellipses; when in addition we find that sediment, several miles in thickness, has been during millions of years deposited and sunken within the central parts of the ellipses, we are ready to render a verdict. Leaving the elliptical theory of the currents entirely out of the question, there are strong reasons for *supposing* that the weight of sediment may produce depressions, and that these by reaction may produce elevations; but without that theory it would be impossible to prove the supposition to be true; and, even if it could be proved, it would be of scarcely any use, either to geology or to geography. So far as utility is concerned, we might as well have the cooling and shrinking theory. The real and vital question is not merely what raised the continents, but what raised them with such forms and analogies, in such positions and number and limits, with such inequalities of surface, and such loxodromic shores and mountains. If it could be positively demonstrated that the weight of sediment sinks ocean basins and raises lands, none of these questions could be answered without the elliptical theory. When the author, in his public lectures in New England and New York in 1853–57, first proposed the geonomic theory, he was not aware that any one had previously suggested that possibly the crust of the earth had, in some places, subsided beneath the weight of sediment. Capt. John Ericsson, of New York, the celebrated engineer and inventor, will doubtless recollect that about that time we discussed the question in his office, without knowing that Herschel had proposed this theory in a

letter to Lyell. It seems, however, that the arguments
of Herschel failed to convince that eminent geologist.
The author has invariably found that scienticians have
regarded the idea of the sinking of the crust beneath
the weight of sediment as a mere speculation, until they
examined the elliptical theory of the currents, and perceived its relation to the forms and positions of the continents.

By means of soundings the depressions and elevations
in the floor of the North Atlantic are well known to
geographers. The greatest elevations are in the middle,
and the depressions not far from the borders of the continents. The most reasonable explanation of this fact
appears to be that, although the greatest depressions
were doubtless in the middle of the basins when the
water covered the whole globe, they became greatest
near the continents when the land furnished coarse and
heavy sediment, which did not reach the centres of the
basins. The volcanoes that rise up above the surface
from the bottom of the ocean probably owe their elevation to the inequalities of the pressures produced by
sediment in different parts of the ocean floor.*

---

* Prof. R. Owen, of New Harmony, Indiana, several years ago
called attention to the fact that the great shore lines of the continents are tangents from the Arctic or from the Antarctic Circle.
No reason for this fact has heretofore been given, but our elliptical theory furnishes the reason, by showing that the continents were
raised in consequence of the sinking of the basins, and that these were
produced by ellipses that were so limited that they could not extend
the shore lines north or south beyond the forty-fifth parallels.

That the theory first published by the author in 1857
has been gradually gaining ground, so far as it relates to
the sources and the depressing effects of sediment, the
following extracts render manifest.

The following is from an English scientific journal
entitled "Nature":

"THE SINKING OF THE EARTH'S CRUST.

"The extreme sensitiveness of the earth's crust to
any changes in the distribution of weight upon its surface is best exemplified by those local depositions of
matter which have attracted general attention at the
present day. The chief of these is the transfer of
matter by river action to large tracts, and its accumula-
tion in such limited areas as plains, estuaries, and deltas.
Borings of four hundred to five hundred feet have
shown that these often consist of long successions of
silts, which alternating layers of shells and of vegetable matter proved to have been deposited at or near the
sea-level, and the wealden and eocene formations in the
British area show that such accumulations may exceed
one thousand feet in thickness. In the case of deltas,
subsidence must keep pace almost foot by foot with the
accumulations, and be confined to the area over which
the sediment is being deposited, for any more rapid
subsidence would check its growth and convert it into
an estuary. This sinking is apparently of universal
occurrence. A similar instance of the transfer of
weight from larger areas and its precipitation on a very
circumscribed area is seen in coral atolls and reefs.

The explanation of their formation given by Darwin requires a gradual subsidence keeping pace with their growth, which takes place within twenty fathoms of the surface only. This theory, simple and admirable as it is, accounting satisfactorily for all the observed phenomena of coral growth, has been contested by Mr. Murray, who has shown that atolls might be merely incrustations of volcanic peaks. But his theory seems improbable by contrast, for it demands two hundred and ninety volcanic peaks at the sea-level in the Pacific coral area alone, every foot of which has been completely concealed by coral growth, though few volcanic craters are known so near the sea-level outside this area. We seem thus to have in coral growths another evidence of subsidence keeping pace with the increase of weight, sometimes, as soundings prove, to a depth of one thousand feet or more. The replacement of a column of sea water one hundred fathoms in depth by a column of limestone, would increase the pressure per fathom from six hundred and nineteen and one-half tons to fourteen hundred and eighty-seven tons, so that it is easy to realize how vast must be the increased pressure on such an area as that occupied by the great reef of Australia, twelve hundred and fifty miles long and ten to ninety miles broad. The sands, gravels, and clays, with marine shells and erratic boulders, prove that a great submergence took place during the glacial period, while Europe was under an ice-sheet six thousand feet thick in Norway, and diminishing to fifteen hundred in Central Germany. The extent of the submergence has been

perhaps understated at six hundred feet in Scandinavia,
and was at least thirteen hundred and fifty feet in Wales.
A corresponding re-elevation accompanied the disappear-
ance of the ice. It has often been supposed that the
sinking of the west coast of Greenland is similarly due
to its ice-cap."

Extract from the Cyclopædia Britannica.

"The bed of the ocean supported on a yielding
substratum may be depressed, without a corresponding
depression of its surface, by the simple laying on of
material, whether abraded from the land or chemically
abstracted from the sea itself. The matter *is* in process
of abrasion and transportation from the land into the
ocean at every instant and along every coast line. We
know too that all existing strata, however enormous
the thickness, *have* been formed at the bottom of the
sea, and it is therefore no hypothesis, but a perfectly
legitimate assumption, that the same process is still in
progress, no matter how slowly, from this cause, at least
in the vicinity of coast lines ; and when we look at the
vast amount of exuviæ which constitutes so large a
portion of the secondary and tertiary beds,—the secre-
tions of mollusca, infusoria, and zoophites,—and bearing
in mind the large proportion of continental substance
which has been so formed, look at the evidence afforded
by deep-sea soundings and by coral formations, that the
same process is still going forward in open sea far out
of the reach of coast washings and river deposits, we
shall at once perceive that any amount of pressure on

the one hand and relief on the other, which the geologist (or geonomer?) can possibly require to work out his
problem, and any law of distribution of that relief and
that pressure is available without calling in the aid of
unknown causes."*

---

* " Dr. Carpenter computes the progeny of a pair of aphides, if
allowed to accumulate, at the end of one year at a trillion. Granting the reproduction of marine animalcules to be a thousand times
less rapid than that of the aphides, granting that each of them
during its lifetime secretes only a ten-millionth part of a cubic
inch of indestructible calcareous matter, we should find accumulated in less than a quarter of a century a globe whose diameter
would exceed the distance travelled by light since four thousand
years before Christ."—*Cycl. Brit.*

SECTION XI.

THE NORTHERN GLACIAL EPOCH.

THERE is no part of dynamical geology for which the experts in that science deserve as much credit as for their persevering researches relating to the Glacial Epoch. The results of their labors may be briefly stated as follows:

At the close of the tertiary age,* and about the time when it is supposed that man was making his first appearance upon the terrestrial stage, the dry lands near the Arctic Sea had risen not less than a thousand feet higher than they are now; and at the same time the climate became excessively cold, even in the temperate zone. The snow accumulated until it was several thousand feet high, and, in the form of glaciers, advanced from the north, and from every high mountain, slowly over half of Europe and North America. All vegetable and animal life was extinguished or driven south, excepting

---

* "In the Glacial Epoch there was an upward movement in high northern latitudes one thousand to two thousand feet, and a change to a colder climate; in the Champlain Epoch there was a downward movement five hundred to one thousand feet below the present level and a moderation of climate; in the Terrace Epoch there was a gradual rising to the present condition."—*Dana.*

such as was adapted to an Arctic climate. How long this condition continued is unknown, but it was probably thousands of years. Then followed the drift period; the lands that had risen subsided, so that they were several hundred feet below their present level, and the climate underwent a change, and became warmer even than it is now; the glaciers melted, and the Arctic Sea vomited forth a large portion of its contents, scattering them upon the lands of the temperate zone. A tremendous flood rushed southward, bearing ice, gravel, and boulders in such immense quantities that it would seem that there could be little left. Then another change occurred: the lands rose again, though not to their former height, but only to their present positions.

The evidence in favor of the facts just stated is of such a nature as to leave no doubt in the minds of geologists of its perfect accuracy. If any of our readers wish for more knowledge of details on this, or indeed of any part of American geology, we would refer them to that masterly work of Prof. J. D. Dana,—the "Manual of Geology."

Various hypotheses have been proposed to account for these wonderful occurrences, some of which seem to be in a high degree visionary, and all have one serious defect,—they do not give any reason for the great elevation of the northern lands; yet this is the most important fact to be explained. Neither the cooling of the globe, the passage of the solar system through an intensely cold region of space, the change of the axis of the earth, the change of its centre of gravity, nor the variations of its orbit will reasonably account for the rising and sinking and re-elevation of the circumpolar lands, and the corresponding changes of climate; still less do these ingenious hypotheses account for the more extreme changes in the southern hemisphere. Prof. Le Conte, after discussing the subject, exclaims, "This is confessedly the most difficult question in geology." Geonomy furnishes the long desired key to the problem.

The Glacial Epoch was the necessary result of the operation of the elliptical currents. They were limited in their range to the zone of the globe between the fifth parallel and the forty-fifth, which is only half-way to the pole. There was therefore a space of forty-five degrees width between each ellipse and the pole. If we assume the edge of the ocean basin to be ten degrees nearer the pole than the northern side of the ellipse, there are yet thirty-five degrees between the edge of the basin and the pole. The sinking of all three of the northern basins, and the elevation of their northern edges above the surface of the ocean, would entirely land-lock the circumpolar sea, cut off its communications with the warm oceans south of it, and prevent the escape of its chilled waters. Under these circumstances, what would happen? The clouds from the south would not be excluded; they would rush poleward, loaded with vapors, which would fall in the form of snow in vast quantities. That which fell in winter would not half of it be melted in summer; it would therefore accumulate more and more; the heaps would rise higher and higher, until glaciers, several thousand feet thick, would

move southward, and increase the width of the frigid
zone by adding to it half of the north temperate zone,
covering its hills with moving glaciers. In a word, there
would be a Glacial Epoch. At length the weight of this
accumulated mass of compact snow and ice would cause
the crust to sink,—then would follow the " Drift Epoch :"
the barriers would give way, the imprisoned arctic
waters would rush forth, the snow and ice would melt
and increase the flood, the high northern lands and
mountains would be torn and abraded, and their materials swept southward over the temperate zone. Then
the northern lands, being thus relieved of their burdens,
would rise again, though not to their former height, for
they would be loaded with the materials of the drift.
There are several reasons why the high northern lands
would in the earlier ages become more elevated than
those in the tropics: 1. In the tropics the elliptical
currents flow quite to the equator, and by depositing
sediment prevent lands from rising there; whereas those
currents do not flow within forty-five degrees of the poles.
2. There are three sinking ocean basins that combine to
force lava under the circumpolar crust; and as the me-
ridian lines come to a point at the pole, the space is narrower east and west. 3. In all other places the sinking
ocean basins have other sinking basins that antagonize
them, but here there is no special opposition.

Although the Glacial Epoch did not commence until
the close of the tertiary age, the elevation of the high
northern lands had been in progress during all the previous geological ages. Indeed, notwithstanding the

8*

check that this upward progress received in the Glacial
and Drift Epochs, it is a question whether the same
causes are not "now in operation," and tending to a
repetition of the same terrible effects.

Originally there were three channels through which
the three great oceans supplied warm water to the
Arctic Sea,—one was through Behring's Strait,* which
was then very wide and deep; a second was through
Siberia from the Arabian Sea; a third was through the
present Iceland Sea. It was the closing of these channels by the upheaval of their beds that produced the
Glacial Epoch. The bursting of the barriers let loose
the imprisoned Arctic waters, and the channels thus
made have since remained open. Let them be again
closed, and, of course, the same effects would follow.
Are there any indications that the high northern lands
are again rising?

Reclus says, " We must not lose sight of the fact that
the upheaval of the north of Scandinavia is not an isolated event, and that other countries of the north of
Europe and Asia all appear to be actuated by a similar movement of ascension. Spitzbergen is gradually
emerging; the northern coast of Russia and Siberia are
likewise rising. It is very probable that the upheaval

---

* " Narrow and shallow as Behring's Strait is,—thirty miles its
narrowest and twenty-five fathoms its deepest part,—it allows a
portion of the circulating water from a warm region to find its
way into the polar basin, aiding thereby to prevent, in all probability, a continual accretion of ice, which else might rise to a
mountainous height."—*Cyclopædia Britannica.*

is prolonged over a great portion of North America. The cliffs of Scotland present phenomena similar to those of Scandinavia.

Dr. Southhall states, that "in the seas about Nova Zembla the sea-bottom has risen one hundred and ten feet in three hundred years. The rise of land in Norway since the Glacial Epoch is still more remarkable."

Mr. Hauworth, in the report of the British Association, says, "The bed of the Arctic Sea is rapidly rising and gaining on the sea along the whole coast line."

Again, he says, "We must seek for the cause of the Siberian change of climate in the draining and elevation of the bottom of the Mediterranean Sea, which once extended from the Euxine to the Kingan Mountains."

The analogies of the three pairs of continents are made very striking if on a common map of the world we restore the North Indian Ocean; but they become still more so if we restore to Europe the territory which rightfully and naturally belongs to her on the northwest. If Alaska and the land as far west as Behring's Strait belongs to North America, then by analogy Iceland and Greenland belong to Europe.

Reclus remarks, that "geologists have established the fact of the former existence of land joining England and Ireland, Ireland and Spain, and even Europe and America."

The theory seems to be by no means unreasonable that during the Glacial and Drift Epochs the Arctic Sea, in bursting out of its bounds, swept away the lands that formerly connected the two continents.

SECTION XII.

THE SOUTHERN GLACIAL EPOCH.

THE geological evidence that there was a Glacial
Epoch in the southern hemisphere is of an uncertain
character, and the whole of the south circumpolar
region is a *terra incognita*. But geonomy indicates, in
the most positive and decisive manner, not only that
there was a south Glacial Epoch, but that it was much
more terrible, extensive, and disastrous than the one
which occurred at the north.

One of the most remarkable facts in geography is
that the greater part of the southern hemisphere is covered with water, while much the greater part of the dry
land in the world is in the northern hemisphere. No
one seems to have had the curiosity to inquire *why* this
is so. We have been contented with the fact, and assumed that it was always thus; but geonomy points to
a different conclusion. If we look at the ideal map, we
see that the original plan, so to speak, was to have just
the same quantity and the same forms of land and of
oceans in one hemisphere as in the other.

The map also indicates the forms, the number, and
the positions of the continents which would now exist
if the elevations had been equal, and if the ideal plan

had not been interfered with by any extraneous causes. Our theory logically compels us to assume that, in the beginning, the continents arose, or began to rise, in accordance with this ideal; and that we must now endeavor to discover, by the best evidence we can obtain, the causes of the departures of the present actual map of the world from the standard or ideal map. Before the land rose high enough to produce a Glacial Epoch there is no known reason why the continents should not have risen equally, or why there should not be as much land in the southern hemisphere as in the northern. But when the Glacial Epoch arrived, it introduced a true and sufficient cause for almost any degree of disturbance and derangement.

Although geology does not enable us to give a detailed history of the Glacial Epoch in the south Polar region, it combines with the elliptical theory to show indirectly what, in all probability, was the course of events in the south by its analogies to those in the north. In the north there was a period of elevation, followed by a period of depression, and then succeeded a period of re-elevation. This is what we learn from geology. From geonomy we learn that there were three southern continents analogous to North America, and that they projected from the Antarctic Circle, just as North America does from the Arctic. The regular and normal sinking of the ocean basins, and the rising of the lands, would cause the exclusion of the warm currents, and produce a Glacial Epoch with all its attendants and consequences.

In the north, after the drift period, the land that had been depressed rose again, but in the south it seems that it still remains in its depressed condition ; in other words, the re-elevating, or "terrace" period, has not yet begun in the south.

What knowledge we now possess of this southern region is fragmental; but geonomy points out a new method of investigation, which, in the hands of future explorers, will, we hope, lead to more satisfactory results. At present we are obliged to feel our way with what means we have.

When we compare the actual map of the world with the ideal, we perceive that the pointed form of South Africa, and also of South America, is abnormal. Those southern points are evidently the remains of broad continents that were wrecked during the Glacial and Drift Epoch. The map indicates that the force which acted against the eastern sides of the southern points came from the southeast. This is the very course—northwest—which a flood or a current would take if it came from the Antarctic coast.

Australia and the islands around it seem to be the shattered and scattered fragments of a continent that was once of unsurpassed magnitude.

It has been asserted by several authors that the plan of the continents and oceans was laid out, and their positions and forms determined, before the continents began to rise. But this is only true of the geonomic or elliptical plan. It is not true that the original *plan* was to have two and a half pairs of oceans,—to have the North

Indian basin raised above the sea and added to the continents, nor to have the North Atlantic so narrow and
the Pacific so much wider.

It is a question not easily answered whether the
North Indian basin would have been elevated and
ruined if there had been no south Glacial Epoch.
We may safely assume that the Glacial Epoch resulted from the land-locking of the circumpolar seas,
and that the "*drift*" and flood was produced by the
bursting of the barriers around those seas. It is probable that the principal bursting occurred on one side of
the pole, and not equally all around it. In the north it
was through the Atlantic that the flood was the most
violent, for it is between America and Europe that the
channels to the Arctic exist, which were doubtless made
during the drift by the passage of the flood. In the south
the condition of the three continents indicates that the
barrier burst near Australia and the flood rushed northwest against the east of Africa. This conjecture is sustained by the fact that the southern point of Africa is
now less extended southward than is South America or
Australia, and also by the fact that the North Indian is
so deep and so destitute of islands. It is opposite this
ocean that the North Indian basin is most elevated ;
and if we assume that the elevation was caused by the
sinking of the South Indian, we may also assume that
it had received the most sediment from the southern
drift.

Independent of geonomy, the observations of Darwin,
and more especially those of Prof. Dana, who made a

personal examination of the southern islands, leave us
in no doubt that a continent, larger than America, is
now gradually sinking, and has been for many centuries. Long ranges of islands are there, which we
have reason to believe are the tops of mountains. According to our theory, mountain ranges are never created
beneath the sea at a distance from any shore. These
mountains, therefore, formerly existed on the border of
a continent that has disappeared.

We must not extend our speculations on this interesting subject beyond the jurisdiction of the facts, but we
cannot help allowing our imagination to carry us back
to the time when several continents existed in the midst
of the southern oceans, and to conjecture concerning the
character and the fate of their inhabitants. Did our
white race originate there, and during the Glacial Epoch
emigrate to Asia?   Who knows?

SECTION XIII.

MOUNTAIN UPHEAVALS.

WHEN an area of land first rose above the sea it was attacked by the winds and waves, and the coarser sediment, after being borne a short distance, was deposited in a line parallel with the shore. The weight of this sediment created local sinking basins, which, by reaction, raised, or tended to raise, parallel ridges. * There is a long submarine slope between almost every ocean shore and the precipitous edge of the continent, and it is on this slope that continental islands are raised. The island is generally curved, and is convex toward the ocean and concave toward the mainland, and also toward the local basin the sinking of which raised it. If the whole submarine slope is afterwards elevated above the sea, the island becomes a mountain and the basin a dry valley.

It is a favorite theory of some writers that all moun-

---

* "About the continents there is often a region of shallow depths which is only the submerged border of the continent. On the North American coast, off New Jersey, it extends out eighty miles with a depth there of only six hundred feet, and from this line the ocean basin dips off at a steep angle "—*Dana.*

E   9

tain ranges were raised on the borders of continents, in consequence of the "lateral pressure" of the sides of the sinking ocean basins; but there are many mountains in Europe and Asia that are so situated, and with such trends, that no oceanic depression could possibly have raised them. Besides, none of the mountain ranges were elevated while the area they occupy was covered by a *deep* sea; the continent rose in the ocean between two and three miles without lateral pressure producing any mountain ranges on its borders, and then, when it was only covered by comparatively shallow water, the mountains *began* to rise.

A perfect illustration of the theory we advocate is furnished by the elevation of the ridge called the "Ulla Bund, or Mound of God." A large area of the coast of India, covered by shallow water, suddenly sank, though not to a great depth, and, parallel with this depression, on the land a ridge about twelve feet high and fifty miles long was raised.

The Rev. W. R. Coovert states that in building a railroad near his residence in Pennsylvania it became necessary to fill up a large mud slough, and the heavy materials that were put in sank in such quantities as to produce astonishment. At length it was discovered that at a considerable distance a mound had risen in the bed of a stream sufficient to turn it out of its course. In this instance we have illustrated on a small scale the process by which mountains are upheaved.

The advocates of the lateral pressure theory find it necessary to state that mountains, as a general rule, are on the borders of continents; it would be more consistent with the facts to say that they were raised on the borders of seas, whether these seas were oceanic or inland. It may be safely laid down as a general rule, that while continents and plateaus were elevated in consequence of oceanic depressions, mountains resulted from local and limited depressions. The curves of mountains and of continental islands bear strong testimony on this subject; indeed, to a dynamical engineer, we should suppose that the evidence would amount to positive proof. He has only to look at a map which represents the eastern part of Asia to see that the mountains there must formerly have been continental islands; they are nearly all curved, and have dry basins on their concave sides. The Himalaya are concave northward toward the great basin of Thibet, the Alps are concave southward toward the basin of Piedmont, and the Carpathians toward the Hungarian basin. In Western America, the Rocky Mountains are concave westward toward the Great American Basin, and the Nevada and Coast Range are concave eastward toward the same basin. Where the shore of a continent was originally long and straight, the sedimentary deposits derived from it would naturally be in a straight line, and the resulting mountains also. But if the shores were indented and irregular, the islands would be more separated, and would be curved. For similar reasons, we observe that the mountains and islands in archipelagoes are curved. The Antilles, the Aleutians, and the East India Islands illustrate this statement.

The difference in the slopes of continents and of mountains indicates the difference in the extent of the basins, the depression of which produced them. The mountains have much steeper slopes than the continents, and, like volcanoes, indicate narrower or more precipitous depression. It is impossible to conceive any movement of the *ocean* basins that could have raised the Alps, but it is easy to conceive that they might be raised by a depression of the basin of Piedmont, when it was a part of the Adriatic Sea.

In order to understand the origin of mountains we must begin by appreciating the fact that if the six ocean basins had sunk equally, and the continents had all risen equally, and all parts of each continent simultaneously, there would have been no mountains except on the borders of continents. All the shores of the continents would have had long submarine slopes; on these slopes sediment, derived from the shores, would have been deposited, the weight of which would have produced depressions, and these, by reaction, would have produced continental islands. When the slope afterwards rose above the sea, the islands would have been ranges of mountains on the borders of the continents. By this proceeding a new shore would be produced, and a new and parallel chain of continental islands, which, when this new slope was raised, would be a new and parallel range of mountains. We have here in a few words given the reason why there are so many ranges of mountains on the borders of continents, and why in many places the ranges or ridges are parallel, those nearest the shore being the last created. We are also enabled to perceive that parallel ranges of upheaved mountains can only be created on a *sloping* submarine area,—for on a level area there could not be the requisite succession of parallel shores.

From the preceding considerations, we infer that any mountains that are not on the borders of continents have resulted from some departure from the ideal,—some irregularity in the rising of the continent. To illustrate, suppose a continent like Australia had risen above the ocean on the western side, while the eastern side remained deep below, so that there would be a long slope downward; then if the land should slowly rise, and advance from west to east, there would be a succession of parallel mountains and valleys over the whole continent. But it seems that both sides of Australia rose at nearly the same time, and there were not shores in the interior a long time enough to create mountains. South America is a triangular continent and has mountains on all three sides, with only low hills in the interior. North America is less favorably situated ; the eastern side rose earlier than the western. Two great interior basins were created : one between the Appalachians and the Rockies, and the other between the Rockies and the Nevada and Coast Ranges. These were both sinking basins that contended long, but unsuccessfully, against the upraising power of the great oceans. The sinking of the interior basins raised the Rockies and the Nevada, and the sinking of the North Pacific raised the western half of the continent.

Southern Africa has its principal mountains on its
borders, and on that account it belongs in the same class
as North and South America and Australia. All four
of these continents approximate to the ideal; but Northern Africa, Europe, and Asia have been rendered abnormal by the uprising of the floor of the North Indian
Ocean; they must therefore receive an especial explanation in another place.

There are three species of mountains: 1. Those created
by the erosion and sweeping away of the surrounding
land, leaving one area isolated in its original position.
2. Upheavals, which have already been described and
accounted for. 3. Corrugations; some parts of the Appalachian Mountains afford perfect illustrations of this
species of mountains. We learn from geology that the
area now occupied by a large section of these mountains
was at first a low and level plain, on which, for unknown
centuries, the coal-producing plants grew in abundance.
Then a force from the east, and doubtless from under
the Atlantic, acted upon this level area and produced its
present corrugated condition.

An examination shows that the force acted very slowly,
and that it produced its most disturbing effects near the
Atlantic, making a succession of ridges with intervening
valleys, the steeper side of the ridges being toward the
west; as we proceed from the Atlantic westward, the in-
tervales or valleys between the ridges become wider and
the ridges less and less elevated.

Two theories have been advanced to account for these
effects. One is that the side of the Atlantic basin, by

lateral pressure,* crowded the continent westward with such force as to produce a corrugation of the strata, which was more pronounced the nearer it was to the ocean. Another is that waves of fluid lava beneath the ocean and the crust were forced under the continent westward. This last theory agrees with that advocated in these pages, which is that all the continents were raised in consequence of the upward pressure of waves or streams of lava from beneath the sinking oceans.

A distinction is. made by geographers between low plains, plateaus, and mountains. Plateaus are high plains situated not far from the ocean. The waves of lava that move obliquely upward from beneath the ocean reach the border of the continent first, and produce their most elevating effects there; they then turn and move more nearly horizontally toward the interior: accordingly, the highest plains or plateaus of a continent are near the ocean, and the lowest lands are distant from it.

It is not difficult to understand that in some instances the waves of lava, in raising a plateau, may have corrugated the strata above them. Prof. Rogers states that he found parallel ranges of folded strata in the Malverns and the Ural similar to those in the Appalachians.

---

* Referring to the force that raised the Appalachian Mountains, Prof. Dana says, "It acted at right angles to the general direction of the Atlantic coast, the flexures being approximately parallel to the coast line. . . . Being greatest on the ocean side (of the continent) and fading out toward the interior. The force was slow in action and long continued—a few feet in a century—without obliterating, and scarcely obscuring, the stratification."

It may not therefore be an unreasonable hypothesis that the Appalachians are a peculiar species of corrugated plateaus.

The theory that the continents were raised by streams of fluid or plastic lava, forced under them from beneath the sinking ocean basins, is strongly confirmed, if not proved, by the relative heights of the lands as indicated by the profiles of the continents. In every one of these profiles we see that the high table-lands or plateaus are near the ocean, and the lowlands at a distance from it.* We are thus told, in language plainer than words, that the liquid matter from beneath the oceans was forced under the continents, and that as the distance from the ocean increased the force and the quantity of the matter decreased. The highest and most extensive plateau in the world is that in Southern Asia, opposite the Indian Ocean. This great sinking basin combined with the North Pacific to raise two-thirds of Asia into a plateau.

---

* " The Andes have been rising century after century at the rate of several feet, while the pampas (vast plains) east of them have been raised only a few inches."—*Lyell.*

SECTION XIV.

THE NORTH INDIAN AREA.

THE elliptical current that formerly circulated in the North Indian was limited, as all other ellipses were, to the space between the equator and the forty-fifth parallel, and the ocean reached but little farther north of that latitude. The ellipse extended westward over the greater part of North Africa, and eastward as far as India and China. The analogies of the three pairs of continents have often been referred to by geographers, but they will be much more remarkable if, on a map of the world, we restore this ancient and now defunct ocean. The area that it formerly occupied is in every respect exceptional.

1. It is the only one of the six great primitive oceans that has had its bed converted to dry land and its elliptical current excluded.

2. It contains much the largest inland salt seas in the world. The Mediterranean, the Euxine, the Caspian, and the Aral, a short time ago, were all united, and constituted a long, narrow, east and west sea, the remains of an ocean once larger than the North Atlantic.

3. Judging by analogy to the other oceans, the ellip-

tical current of the North Indian flowed westward, near
and parallel to the equator, and, turning northwest,
it continued to the western extremity of North Africa
(which was then a gulf similar to the Gulf of Mexico);
it then turned northeast through the Mediterranean
and Caspian, then south to the Bay of Bengal, thus
completing its circuit. Being analogous to the other
ellipses, it doubtless sent an offset or warm branch
northeast through Siberia to the Arctic Sea. The
Arabian Sea, which was then a part of the North Indian, is now one of the warmest maritime places in the
world; not long since it supplied warmth to the current
that flowed through the vast and then salubrious Siberian plains, where hordes of elephants and other large
animals found a genial climate and plenty of food.
The elevations of the land, by excluding the warm current, proved fatal to the elephants, whose ivory remains
still testify to their former existence and their great
numbers.

4. The North Indian area contains the most extensive tertiary formations* in the world. These testify

---

* "The occurrence of Numelitic limestone in the Himalaya
Mountains, at the height of sixteen thousand feet above the sea,
shows that this great range has been lifted above the sea sixteen
thousand feet since the Eocene Seas covered Thibet and Central
Asia."—*Southhall.*

"The tertiary extends from the Baltic to the Black Sea, and
including nearly the whole of Belgium, the Netherlands, Den-
mark, Hanover, Prussia, Poland, great part of Austria, the south
provinces of European Russia down to the Black Sea, the whole
of the Caucasian district between the Sea of Azof and the Caspian,

that but a short time ago, geologically speaking, it was covered by an ocean, while many of the surrounding places were dry land.

5. This area contains the highest and most extensive plateaus in the world. Those plateaus were evidently raised by the lava forced under them from beneath the sinking South Indian Ocean.

6. It contains the highest and most numerous mountains, in proportion to its extent, of any place in the world.

7. It contains the lowest dry lands in the world : the Sahara, the vicinity of the Caspian and of the Dead Sea, are below the level of the ocean.

8. It contains the only mountains of importance that trend east and west. The Scandinavians and the Urals that trend north and south were not within the bounds of the North Indian ellipse, neither were the mountains in the extreme east of Asia; but the Atlas, the Alps, the Caucasus, the Carpathians, Thian-Shan, Kuenlun,

---

nearly the whole of Western Tartary, nearly the whole of Siberia, the great desert of Africa, as well as the bulk of Arabia, Persia, and Upper India, the deserts of Shamo and Gobi." All these countries are tertiary.—*Ency. Brit.*

"Between the Himalaya and the Kuenlun the plateau of Thibet is an oval expanse about five hundred and fifty geographical miles across. It is the most elevated area of level ground on the globe. It forms the southernmost of the three great table-lands of Central Asia.

"Gobi, like Shamo, means a sandy desert. 'Gobi is a dry sea, suggestive,' as Forsyth says, 'not only of its present appearance, but also of its former history. As a sea, Gobi must have been comparable to the Mediterranean, and the ancient coast line can be pretty clearly recognized."—*Ency. Brit.*

Hindoo Kosh, and Himalaya are within those bounds. Geographers have often remarked the great difference between North and South Africa, but none of them have perceived the reason of it. Prof. Guyot, with his usual sagacity, observed that Northern Africa appeared to belong more to Europe than to South Africa; and Prof. Dana has pointed to the fact that Northern Africa, though occupied by dry land, is analogous to the Gulf of Mexico,* which is covered by water. The truth is, that it *was* formerly a gulf like that of Mexico, and had an analogous Gulf Stream circulating within it, the western *extremity* of Africa at that time being dry land, and analogous to the present isthmus of Central America.

In order to understand the natural history of the North Indian area, we must conceive of it as at first an ocean, like the North Atlantic and the North Pacific, with a similar current to the Gulf Stream, flowing in it—with the continent of Europe, partly emerged, on its western side, Asia struggling upward from the sea on the east, and South Africa, mostly dry land, on the south. We must next conceive this great basin as being gradually undermined and rendered less deep by the upward pressure of lava forced under it from beneath the North Pacific, the South Indian, and the South Atlantic, until, in the tertiary age, it was transformed into a vast archipelago,—a shallow sea, studded with islands and peninsulas, and surrounded by shores

---

* " The western expansion of Africa corresponds to the indentation of the Caribbean Sea and the Gulf of Mexico."—*Dana.*

that were advancing in all directions toward its centres and thus contracting its area. Next we must conceive of its elliptical current,—its "Gulf Stream,"—after many a struggle and change of its course, entirely excluded. What then remained of this doomed ocean was a long, narrow, Mediterranean Sea, that extended eastward from the western extremity of Europe and North Africa to the eastern parts of China. It was the east and west shores of this sea that gave birth to the east and west mountains of Asia, Europe, and North Africa during the tertiary age. If the Indian Ocean had maintained its original condition, there would have been no east and west ranges of mountains in Asia nor in Europe. Our knowledge of the geology of the Old World is very imperfect, but it is certain that the Scandinavians, the Urals, and the Altai were in existence when the Himalaya and the Alps and all the other east and west mountains were yet covered by the ocean.

The ground plan of Asia and Europe was laid out, —the positions of its mountains and valleys arranged, while it was yet an archipelago. The basins of Thibet and Shamo and Gobi, of Hungary and Piedmont and Switzerland, the plains of Arabia and Siberia and Sahara, were all covered by a tertiary sea when the present mountains were low islands amid those seas, or just beginning to emerge from them. The sinking of the local basins raised the mountains; but it was the sinking of the great *ocean* basins that raised the whole archipelago, and transformed its basins into dry valleys.

Although the mountains were first raised by the de-

pression of local basins, it is reasonable to suppose that
the force that elevated the whole archipelago would act
specially and raise the mountains more than they would
the more heavily-loaded basins. This inference is sustained by some observations lately made by geologists.
They have found evidence that in several instances the
tops of mountains have been raised, while their lower
parts have remained undisturbed. It is also sustained
by the fact that the pendulum indicates more thick and
dense masses beneath the surface of the earth in valleys
than upon mountains.

THE CAUSE OF EARTHQUAKES.

When the fluid lava is forced up under the continents
from beneath sinking basins, they produce movements
of the surface. When the movements are perceptible
we denominate them earthquakes. Linnæus was led to
suspect that the north part of Sweden was rising. He
therefore fixed several marks by which future observers
could determine with certainty how much the continent
rose in a given time. After a hundred years had passed,
Sir Charles Lyell found that the land had risen three
feet,—yet the inhabitants had not observed any movements. Sometimes the upheavals are sudden and violent, and the consequences terrible. The probability is
that the sinking of the great and general oceans' beds
raise the continents slowly and imperceptibly, and that
the more limited and local basins, which receive an abundance of sediment from the neighboring shores, produce more rapid and dangerous upheavals. Our

theory is that all elevations of continents and mountains, and all earthquakes and volcanoes, have one general cause, which under different circumstances produces various effects. If the area of depression is very large, we can conceive that the elevation would be broad and continental, and the slopes gradual; but if the depression is narrow and deep, the elevation would be narrow and the slopes steep, as in the case of volcanoes and some mountains. In accordance with this theory we know that while volcanoes burst forth suddenly and upheave a narrow area of land in the form of a cone, the continents have been millions of years rising from their original places at the bottom of the ocean. The results of the explorations and dredgings made at the bottom of the Atlantic prove that in the middle of the ocean the deposits of sediment are very light, while near the continents they are abundant. The smallness of the quantities of sediment deposited in the oceanic centres may, at first, seem to militate against the idea of the upheaval of continents by its weight, but in reality this agrees with the testimony of geology in regard to the long, long time, the almost eternity that has elapsed since the beginning of the creation of the continents. The much more rapid rise of mountains confirms the conclusion that they owe their elevation to the more local and abundant deposits of sediment near the shores. By similar reasoning we are led to the conclusion that the most destructive earthquakes result from comparatively narrow areas of depression, and not from the slow subsidence of the great ocean basins.

SECTION XV.

CONCLUDING REVIEW.

IN reviewing our treatise, let us consider some of the important questions that can be answered by geonomy, and not without it, and in this manner estimate the value of the new system.

1. Why are there three pairs of continents?
2. Why are they so analogous to each other?
3. Why are there only two and a half pairs of oceans?
4. Why are the analogies of the continents all included in the zone between the equator and the forty-fifth parallel?
5. Why are the present elliptical currents all included in the same zone?
6. Why are the trends of so many of the shores loxodromic?
7. Why are the southern oceans and continents placed about fifty degrees of longitude east of the northern?
8. Why are some of the mountain ranges east and west?
9. Why are so many mountains and continental islands concave toward the small local basins, and convex toward the great oceans?

10. Why are South America and South Africa pointed at their southern extremities, and why are the eastern sides also pointed toward the east?

11. Why is there so much land in the northern hemisphere, and so much ocean in the southern?

12. What is the cause of the counter-currents?

13. Why does the ocean current divide in the Atlantic,—one branch flowing due east near Newfoundland, and the other branch northeast to the Arctic?

14. What produced the Glacial Epoch?

15. Why is the position of Northwestern Africa so analogous to that of the Gulf of Mexico?

16. Why is there no land between the oceans near the equator?

The critical reader will perceive, after perusing the preceding pages, that if it is admitted that one of the oceanic ellipses existed when there was no land, the whole system of geonomy, with all its details, necessarily follows; for,—

1. There must a sufficient number have existed to give circulation to the whole ocean, and the same number were required in one hemisphere as in the other; accordingly, in studying the map, we find that there were three of them in each hemisphere.

2. Each ellipse must, from its very nature, have continually accumulated floating and sedimentary material in its central parts, which, however slowly, by its weight, would in time produce a sinking basin, or fill the ocean full; accordingly, we find that three pairs of ocean basins were actually created.

3. Three pairs of such basins could not have sunk without forcing the subjacent fluid or plastic matter into the inter-oceanic spaces, and elevating them, and thus giving birth to three pairs of analogous continents; accordingly, we find that three pairs were elevated.

4. If the three southern ellipses were, by any cause, placed east of the three northern, the continents would be equally unsymmetric, and the tropical parts distorted and loxodromic; accordingly, we find such to be the case.

5. The sinking of the three pairs of basins would necessarily land-lock the circumpolar seas and produce their glaciation; accordingly, we have abundant evidence that there was a Glacial Epoch at the north and another at the south, though they were probably not simultaneous.

6. The circumpolar Glacial and Drift eruptions would inevitably tend to deface and denude the lands in the frigid zone and its neighborhood; accordingly, we find the ideal map in those regions greatly deranged. In the north a large area of Europe, and in the south large portions of South America, South Africa, and Australia, are suppressed.

7. The tendency of an elliptical whirl of the water in an ocean basin is to convey the sediment into the central part to produce the greatest depression there, and, by raising the edges of the basin nearer and nearer to the centre, to narrow the basin, and extend the interoceanic continents; accordingly, every geologist knows that, as a general rule, the continents have, from the beginning,

extended themselves at the expense of the oceanic areas. It is, therefore, not true that the ocean basins have by lateral pressure pushed the continents back and made them narrower.

8. The fluid or plastic matter forced obliquely upward under a continent would naturally raise the border of the continent more than it would the interior; accordingly, we find all the plateaus near the borders, and the lowest lands far in the interior.

9. The sediment, before there was any dry land, must have been chiefly organic and chemical or meteoric, and consequently it accumulated slowly; but when the land arose above the sea, and was subjected to the action of the winds, the waves, and frosts, the sediment became abundant, particularly in the vicinity of the shores and on the long submarine slopes; accordingly, the earliest geological ages were of great length, and the later comparatively short; and, besides, the mountains, which were all raised in consequence of the sinking of local basins near shores, rose much more rapidly than continents.

10. Local basins, being small, would naturally have shorter curves than larger basins; accordingly, we find that many mountains and continental islands, and those in archipelagoes, are generally curved, and are concave toward the small basin.

If a philosophic angel, with a knowledge of the principles of geonomy, could have been seated on some distant world, and have seen our globe when the ocean first covered it, and "the Spirit of God moved upon the face of the waters," in elliptical paths, he could, by mere deductive reasoning, have predicted all the most important events and changes that have since occurred in the physical history of the earth. He could have foreseen that the sediment would accumulate in the centres of the ellipses and produce three pairs of sinking basins, and raise three pairs of analogous continents, and that, consequently, the circumpolar seas would be land-locked and glaciated, and then burst forth and produce terrible floods of water and ice, gravel and boulders.

He would have foreseen that the eastern positions of the southern ellipses would distort the central parts of the continents; that the bottom of one of the basins would be liable to be elevated above the sea, and produce exceptional east and west shores and mountains; and that nearly all the other mountains and plateaus would be raised near the borders of continents.

The present writer had not the advantage enjoyed by the supposed angel. He was under the necessity of reasoning by induction, from the facts and phenomena which the earth now presents, back to the time when the continents were in embryo, and the oceans were "boundless, endless, and sublime." The teachers of physical geography in our schools may now avail themselves of both processes of reasoning. By means of the principles illustrated in the foregoing pages, they can begin by proceeding deductively from the first to the last chapter in the earth's history, and then reverse the process, and demonstrate the truth of the principles by well-established facts relating to the present condition of the earth.